T0214443

SpringerBriefs in Fire

Series editor

James A. Milke, College Park, USA

More information about this series at http://www.springer.com/series/10476

Rosalie Wills · James A. Milke
Sara Royle · Kristin Steranka

Best Practices for Commercial Roof-Mounted Photovoltaic System Installation

 Springer

Rosalie Wills
Department of Fire Protection Engineering
University of Maryland
College Park, MD
USA

Sara Royle
Department of Fire Protection Engineering
University of Maryland
College Park, MD
USA

James A. Milke
Department of Fire Protection Engineering
University of Maryland
College Park, MD
USA

Kristin Steranka
Department of Fire Protection Engineering
University of Maryland
College Park, MD
USA

ISSN 2193-6595   ISSN 2193-6609   (electronic)
SpringerBriefs in Fire
ISBN 978-1-4939-2882-8   ISBN 978-1-4939-2883-5   (eBook)
DOI 10.1007/978-1-4939-2883-5

Library of Congress Control Number: 2015940973

Springer New York Heidelberg Dordrecht London

Printed on acid-free paper

Springer Science+Business Media LLC New York is part of Springer Science+Business Media
(www.springer.com)

Preface

Although PV systems provide many benefits to the environment, there are hazards associated with them being installed onto rooftops. The installation of PV systems on roofs creates electrical, fire, structural, and weather-related hazards that are not adequately addressed by current codes, standards, and guidance documents. Significant progress has been made in the past years (as will be identified in this report), but there are still gaps that need to be addressed.

The purpose of this book is to compile information on a wide variety of hazards and damage potential created by the installation of photovoltaic (PV) systems on commercial roof structures.

The book reviews recent major PV fire incidents including those at Bakersfield, CA, and Mount Holly, NC, Goch, Germany, DeLanco, NJ and LaFarge, WI, and concludes that much can be learned from these and other non-fire-related failure incidents, most of which are not documented in the public literature.

The book then summarizes basic performance categories associated with PV panel installation practice and identifies key installation features impacting this performance. These include performance under structural loading, wind loads, hail, snow, debris accumulation, seismic loads, and fire hazards including flammability of components, ignition hazards, and electrical hazards associated with fire fighter operations.

The book reviews existing information in the literature related to the best practices for installation to address the performance issues described above. A comprehensive reference section is provided.

Finally, an assessment of key gaps in available information and understanding of performance is presented, highlighting areas of additional needed work. These include:

- Long-term performance of PV modules
- Design for wind in the presence of deflectors and shrouds
- Long-term performance with respect to hail damage
- Design for accumulated snow load

Acknowledgments

Appreciation is extended to the Fire Protection Research Foundation for support to this project. Special appreciation is extended to Kathleen Almand and Casey Grant for their assistance in identifying resources and providing comments on the direction of research. The comments and input from the Project Technical Panel helped to direct the study and improve the final report.

Project Technical Panel

Larry Sherwood, Solar America Board for Codes and Standards
Remington Brown, Insurance Institute for Business and Home Safety
Bill Brooks, Solar Energy Industries Association
Tom Smith, TL Smith Consulting, Inc.
Barbara Mentzer, City of Hartford
Daniel Joyeux, Efectis (FR)
Michael Bertels, Dutch Fire Department

Project Sponsors

CNA Insurance
FM Global
Liberty Mutual
Tokio Marine
Travelers Insurance
XL Gaps
Zurich NA

Contents

Chapter 1
Introduction

1.1 Research Objective

The purpose of this literature review is to compile information on a wide variety of hazards and damage potential created by the installation of photovoltaic (PV) systems on commercial roof structures. The hazards addressed are:

- structural loading
- wind loads
- hail
- snow
- debris accumulation
- seismic
- fire (panel flammability, impact on roof fire ratings)
- electrical hazards affecting fire fighting operations.

1.2 Background

The environmental movement from fossil fuels to clean, renewable energy has resulted in an appreciable increase in the use of PV systems for electricity production. Just within the second quarter of 2013, the Solar Energy Industries Association (SEIA), has measured a 15 % increase in megawatts of photovoltaic capacity. By the end of 2013, SEIA is expecting that a solar project will be installed in the United States at an average of every 4 min (SEIA 2013). This growth is also being seen globally. China, India, Japan and Germany all have plans to decrease dependence on fossil fuels and nuclear power, replacing the energy obtained currently from these sources with that from renewable sources (Gandhi 2011).

© Fire Protection Research Foundation 2015
R. Wills et al., *Best Practices for Commercial Roof-Mounted Photovoltaic System Installation*, SpringerBriefs in Fire, DOI 10.1007/978-1-4939-2883-5_1

The upfront cost could deter people from investment in PV systems. "The largest barrier to the proliferation of PV technology is its initial cost, and reducing this cost will further promote its widespread use. This obstacle hinges directly on the manufacturing process used to create the solar cells and related technology components" (Grant 2010). The PV systems industry has addressed this issue by lowering the average cost of a PV system and as a result has increased the popularity of PV systems even further. The average cost of a PV system has dropped by 11 % since 2012 and the average price of PV modules has decreased by 60 % since the beginning of 2011 (SEIA 2013).

The benefits of installing PV systems on roofs are identified by Kirby:

> Low-slope roofs are ideal locations for PV systems: the solar resource is good; power is generated in close proximity to loads; the location is secure and unobtrusive; and one- and two-story buildings in particular have favorable ratios of roof-to-wall area. Best of all, low-slope roofs are plentiful. As a platform for PV systems, they represent an excellent business opportunity for both PV and roofing contractors (Kirby 2011).

Although PV systems provide many benefits to the environment, there are hazards associated with them being installed onto rooftops. The installation of PV systems on roofs creates electrical, fire, structural, and weather-related hazards that are not adequately addressed by current codes, standards and guidance documents. Significant progress has been made in the past years (as will be identified in this report), but there are still gaps that need to be addressed. The hazards can pose a drag to implementation of a technology whose growth is being encouraged globally due to its use of a renewable energy source with a potential to decrease dependence on fossil fuels and nuclear power (Gandhi 2011).

Recent studies have evaluated these hazards and recommend mitigating design or installation strategies. A scientific understanding of these hazards is fundamental to developing guidelines and training techniques. Quantifying those hazards and observing what happens in different situations helps better understand the hazards and develop mitigation strategies. Some of these studies have resulted in resources that have developed best practice guides and training methods for PV system operators.

Although much research has been done on PV systems, there are still limitations on the hazards associated with PV systems, especially related to aging. "The relative newness of rooftop PV systems means that the durability and longevity of certain practices are simply not known" Kirby (2011).

1.3 Overview of Previous Incidents

Hazards of PV systems on roofs have caused several incidents; most notably in Bakersfield, CA, and Mount Holly, NC. These fires and their causes are discussed further in the Fire Hazards section. Although these events were unfortunate, the field grew in knowledge from learning about what happened during those events. Brooks (2012) explains:

"The investigations into these fires expose a "blind spot" in ground-fault protection in larger PV systems and provide an opportunity to explore the safety implications of inadequate ground-fault protection in a public forum. As investigators develop an understanding of the root causes of the Bakersfield and Mount Holly fires, they will also develop a better understanding of the complex nature of faults and fault currents in PV arrays, which will benefit all stakeholders." Grant (2010) also states that these incidents would lead to more strict code requirements. "Although the installation met the requirements of the applicable electrical code, this event indicates a need to revise code requirements to provide emergency responders with appropriate measures to readily isolate solar modules."

These two fires are significant in that they were investigated in depth and their source was determined to be associated with the PV system. These fires have influenced the development of future codes.

Another set of fires have occurred in the past few years and have had PV systems associated with the fire. There is not enough evidence as of yet to determine if the PV system was actually the source of the fire, but these examples do show how more fires are occurring where PV systems are located due to the increase in popularity of PV systems.

In May of 2013 a fire occurred at Organic Valley's corporate headquarters in La Farge, Wisconsin. A photograph from the fire is presented in Fig. 1.1. Preliminary investigation has hypothesized that the construction of the green building influenced the spread of the fire. The wood frame, recycled insulation, and the PV system all contributed to the development of the fire. In this specific case, the entire roof became energized because of the combination of the fire, PV system, and the metal roof. This limited the ability of the fire department from preventing the spread of the fire to other parts of the building (Duval 2013).[1]

Another noteworthy fire occurred at a warehouse in 2012 in Goch, Germany, involving a 4,000 m^2 area. Although further information on this fire is limited, one local fire department article stated that the fire chief determined that the cause of the fire was due to a "technical defect in the photovoltaic system" (Feuerwehr 2012).[2]

A recent fire occurred on September 1st, 2013 at the Dietz and Watson Factory in Delanco, NJ. Much of what is known about this fire comes from news reports; these can not be assumed to be reliant and may be proven inaccurate after further investigation. With that in mind, the following details were received from various news reports. The warehouse was the size of six football fields and had over 7,000 solar modules covering the rooftops (Bayliss 2013). An important note from the fire was stated by an article in *Fire Engineering*, "Officials say the fire was contained between the trusses and solar panels on the roof" (Fire Engineering 2013). Unprotected combustible roofing was said to allow the fire to spread and provided significant fuel source to the fire. News reports stated that the PV systems inhibited the fire fighters' ability to suppress the fire and that it took more than 24 h to bring it

---

[1]This information was received from Bob Duval based on his communication with the Chief Phillip Stittleburg of the La Farge (Wisconsin) Fire Department.

[2]An official determination of the cause of the fire has not yet been released, pending an in depth investigation.

Fig. 1.1 Fire at Organic Valley's corporate headquarters in La Farge, WI, XWOW.com (2013)

under control. The entire building and its contents were completely destroyed during this incident. The building after the fire is displayed in Fig. 1.2 with some of the solar panels along the edges of the roof still being intact. The cause of this fire has not yet been determined but it is an example of how relevant the discussion of PV system hazards can be.

Another major incident occurred on November of 2013 in New Jersey. The building was a 700,000 ft^2 Christmas goods warehouse with over 8,000 panels on the roof. Over 300 panels were involved in the fire and the fire did not enter within the building. An early notification from a passing truck allowed for the fire to be contained by the local fire department (DiSanto 2013). Investigation is still preliminary so the cause of the fire has not been determined.

The Bakersfield and Mount Holly fires were investigated thoroughly and reported in some detail. However, many other investigated incidents have not been reported in any detail. This leads to the assumption that only a few incidents have occurred. However, informal estimates suggest that there are more than a dozen similar events that were not reported (Brooks 2013).

There is also a small amount of loss data on the hazards of PV systems on rooftops. Grant (2012) puts the little loss data that there is into perspective:

Detailed loss information to support each of these scenarios is lacking due to the relative newness of this technology. Traditional fire loss statistics such as NFIRS (National Fire

Fig. 1.2 Dietz and Watson fire aftermath (FOX 2013)

Incident Reporting System) handled by the U.S. Fire Administration and FIDO (Fire Incident Data Organization) administered by the National Fire Protection Association, do not provide the necessary level of detail to distinguish the relatively recent technologies of solar power systems. A preliminary scan of the NFIRS data yields 44 incidents that involve "solar" in some manner, but a detailed review indicates that most are not applicable and involve fires that started with sunlight through glass, landscape lighting, are non-structural fires such as vehicles, vegetation, rubbish, etc. Further, proprietary information may exist with certain insurance companies and similar loss control organizations, but this is typically focused on their specific constituents and transparent data summaries are not known to be readily available.

In summary, statistical data involving solar power systems is not readily available to provide quantifiable data analysis of these systems. We do, however, have quantifiable data on the number of structure fires in the United States each year. For example, in 2007 there were 530,500 structure fires resulting in 3,000 deaths, 15,350 injuries, and $10.6 billion in direct property loss. Of these fires, one- and two-family homes accounted for 399,000 fires, 2,865 deaths, 13,600 injuries, and $7.4 billion in direct property loss. While the actual percentage of overall buildings with solar power systems and those involved with fire remains a quantifiably mystery, we have a general expectation of how the data will likely trend in the future. As solar power systems continue to proliferate, the likelihood of fire fighters encountering them at a structural fire will similarly increase (Grant 2010).

These incidents are alarming and are a direct presentation of why the hazards associated with PV systems need to be understood. These incidents are only associated with fire. There are more incidents that have occurred from other hazards such as wind, hail, snow, and extreme temperatures but there is not much documentation on these incidents. The purpose of this report is to review and summarize studies and the associated literature, as well as identify best practices for commercial roof-mounted PV systems in order to analyze potential gaps in current standards and practices.

1.4 Overview of Photovoltaic (PV) Systems

There are different types of PV systems. They can be distinguished between residential, commercial, and utility-scale systems. They can also be classified as ground-mounted, shade structure, roof-mounted or building-integrated PV (BIPV) arrays. PV systems also have varying power capacities. PV systems can range from having a capacity of less than 10 kW to over 100 kW. PV systems installed on commercial roofs typically have capacities between 10 and 100 kW. This report focuses only on the installation of PV arrays on commercial roofs.

Grant (2010) summarizes the basics of photovoltaics:

> The photovoltaic process converts light to electricity, as indicated by the root words *photo* meaning "light" and *voltaic* meaning "electricity", and often represented by the acronym PV. The process involves no moving parts or fluids, consumes no materials, utilizes solid-state technology, and is completely self-contained. The primary concern for emergency responders with these systems is the presence of electrical components and circuitry that present an electrical shock hazard.
>
> The basic components of a photovoltaic system include the photovoltaic unit that captures the sun's energy, and inverter that converts the electrical power from DC to AC, electrical conduit and other electrical system components, and in some cases a storage battery. At the heart of the system is the unit that is actually capturing the sun's electromagnetic energy in the form of light. Figure 1.3, illustrates the basic photovoltaic components used to capture solar energy.

Grant (2010) continues:

> A typical PV module includes not only the solar cells, but several other important components including the concentrators that focus the sunlight onto the solar cell modules, array frame and associated protective components, electrical connections, and mounting stanchions. Figure 1.4 provides a relatively detailed illustration of the primary components

Fig. 1.3 Basic photovoltaic components used to capture solar energy (Grant 2010)

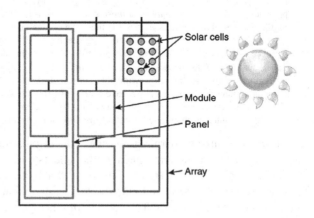

of a PV solar power system, and Fig. 1.4 illustrates the fundamental electrical interrelationship for photovoltaic systems that are stand-alone, hybrid, or interactive with the building's conventional electrical system.

All of these components are designed with significant attention given to their endurance, recognizing that a typical solar panel will be exposed to ongoing harsh weather conditions that will promote degradation. Some of the materials used might have excellent weather endurance characteristics, but not necessarily be resistant to exposure fires. Today, the lifespan of a typical solar array is typically in the 20 to 25 year range, and component endurance is an important performance characteristic of the overall solar energy system.

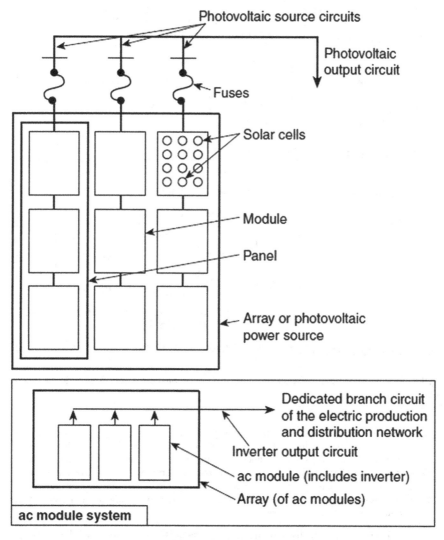

Fig. 1.4 Basic components of a photovoltaic solar power system (article 690 of NFPA 70, 2014 ed.)

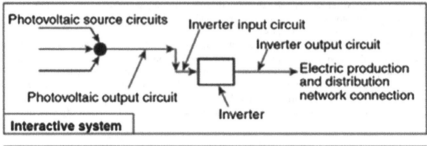

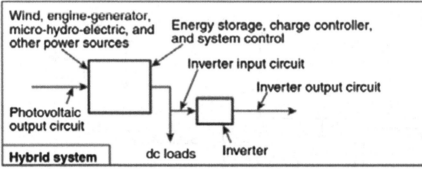

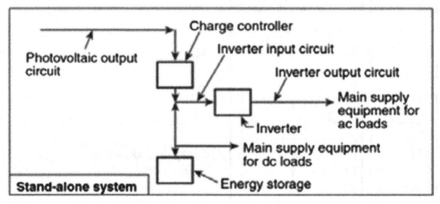

Fig. 1.5 Photovoltaic system interrelationship with conventional electrical systems (article 690 of NFPA 70, 2011 ed.)

In addition to the solar module, the other key components of the PV system are the inverters, disconnects, conduit, and sometimes an electrical storage device (i.e., batteries). The electricity generated by PV modules and solar arrays is dc (direct current), and an inverter is required to convert this to ac (alternating current). As with any electrical equipment that is tied into a building's electrical circuitry, disconnect switches are required for purposes of isolation. Some systems also include batteries to store the additional energy created during sunlight hours for use at a later time (Grant 2010).

The following chapters discuss the selected hazards associated with PV systems and design guidelines proposed to mitigate those hazards (Fig. 1.5).

Chapter 2
Structural Loading

Installing PV panels onto roofs introduces hazards that can affect the structural integrity of the roof. Not only does the roof support the dead load of the PV system itself, but also external forces introduce structural loading. Outside installations exposes the PV system and roof assembly to hazardous elements such as wind, hail, snow, debris, and extreme temperatures. These elements introduce substantial loads to the panels and the roof through wind up-lift, thermal expansion, and debris build-up. Substantial loads can lead to the destruction of rooftops and PV systems. "Structural engineers must consider each of these loads separately and in combination to identify the worst-case loading situation" (O'Brien and Banks 2012). There are guidelines on the installation, maintenance, and testing of PV systems that can help prevent failure of the system due to extreme external forces.

Guidelines depend on what type of mounting is used to attach the PV systems to the roof. There are three different methods of mounting PV systems to a roof structure: ballast-only, attached roof-bearing, and structurally attached. Ballast-only PV systems are weighed down by heavy materials such as concrete to keep them located in the same position. Ballast-only systems are not attached to the roof structure. An attached roof-bearing system uses friction clips to secure PV modules to the beams of the framing system. Structurally attached PV systems are attached to the roof structure such that the load path is the same for both upward and downward forces (SEAOC 2012a). The three types of methods: ballast-only, modular, and structurally attached are shown in Figs. 2.1, 2.2 and 2.3 respectively. Each method has advantages and disadvantages with cost and how different hazards will interact with the system.

There is also another method of attachment, which is BIPV (Building Integrated), these systems are in more green buildings and are becoming more popular. This attachment method is most similar to the attached roof bearing system. Because this attachment method is dependent on the building that the PV system is attached to, diverse building designs leads to diverse BIPV systems. The uniqueness of these systems are displayed in Figs. 2.4 and 2.5.

© Fire Protection Research Foundation 2015
R. Wills et al., *Best Practices for Commercial Roof-Mounted Photovoltaic System Installation*, SpringerBriefs in Fire, DOI 10.1007/978-1-4939-2883-5_2

Fig. 2.1 Ballast-only PV system ASCE (2013)

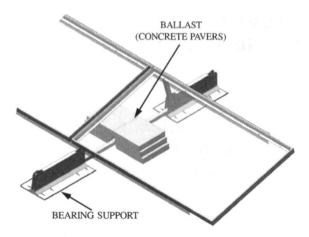

Fig. 2.2 Attached roof-bearing PV system ASCE (2013)

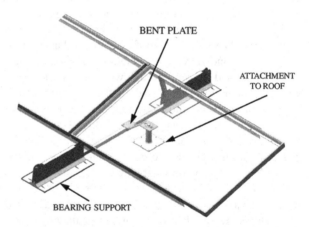

Fig. 2.3 Structurally attached PV system ASCE 2013

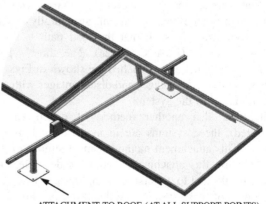

Fig. 2.4 BIPV mounted system (NREL image gallery)

Fig. 2.5 BIPV mounted system (NREL image gallery)

Although these two examples are both BIPV systems they look and behave completely differently. "Building-integrated systems are integral with the roof or lay flat on the roof surface such that they do not affect the roof profile. They may consist of sheets of photovoltaic material attached to the roof membrane by adhesive, for example" (Maffei 2014). The attachment method can significantly affect the loads that are being applied to the structure and how it is being handled. "The roofing industry has learned from experience that ballast-only rooftop equipment does not necessarily remain stationary. Structurally attached equipment is more reliable in this regard" (Kirby 2011). An engineer using calculations found in codes and standards can evaluate structurally attached equipment.

Even though structurally attached equipment can be evaluated by an engineer, the ballast-only systems are difficult to evaluate. There are advantages for ballast-only systems. "Ballast-only systems avoid the cost and increased water intrusion risk caused by the roofing penetrations needed for attached systems" Ward (2013). Ballast-only systems do not puncture the roofing membrane in order to keep them stable, which avoids the issue of water leaking through the holes created by structurally attached equipment. Another aspect of the ballast-only PV systems that creates a hazard for the roofing structure is all the added weight of the ballasting. A roof needs to be able to support not only the PV systems, but also the heavy ballasts that are used to hold down the PV systems. As stated before, the structural loading of the PV systems can be significantly increased when combined with external forces such as wind.

Modular systems are even more difficult to evaluate than ballast-only systems. They are attached to the building by screws, clips, or adhesives. A variation of the attached roof-bearing type is one in which an anchor is used to secure the pedestals of the solar panels to the roof deck as opposed to the roof framing. In that case, the designers should consider the load path for the wind uplift load transferring from the anchor through the deck into the secondary structural framing supporting the deck.

Structural loading is difficult to be calculated for these attachment methods. This method is often used for smaller buildings like residential buildings; therefore this report will focus on ballast-only systems and structurally attached systems.

"It is important that designers and engineers determine loads on modules, fasteners, all components within the racking system and the applied loads to the roof. Loads must ultimately be transferred from the modules to the fasteners and racking system, and ultimately through the roof deck and building structure to the ground. This is common knowledge for most structural engineers. Remember that this likely involves the use of different effective wind areas based on the load- sharing capability of the component under analysis. The wind load rating of the module should not be exceeded. Once wind loads are determined, structural engineers must apply appropriate safety factors and combine loads as required in *ASCE 7-05* Section 2. '*ASCE Standard 7-05* is the standard for evaluating wind forces on structures." The *ASCE Standard 7-05* "provides requirements for general structural design and includes means for determining dead, live, soil, flood, wind, snow, rain, atmospheric ice, and earthquake loads, and their combinations that are suitable for inclusion in building codes and other documents" (Thomson Reuters 2013).

"In addition to wind loads, other loads such as snow, seismic and gravity (dead load) must be taken into account" (O'Brien and Banks 2012). The load that is provided by the weight of the PV systems themselves is only a portion of the loads that is going to be imposed on the roofing structure.

Chapter 3
Wind Loads

An additional complexity to having PV systems on rooftops is that the PV system will be exposed to wind forces and as a result will have to be capable of withstanding those forces. PV systems must withstand escalated weather scenarios such as windstorms. Uplifts from strong winds can create appreciable additional loads or load concentrations. The very presence of the building changes the aerodynamic load because "there is a complex interaction between building generated vortices and the flow induced by the array, which depends on building height, the setback of the array from the roof edge, and other building parameters" (Kopp et al. 2013).

The three types of attachment systems handle external forces in different manners. For ballast-only systems, resistance to wind and seismic forces is provided by weight and friction. For attached roof-bearing systems, the load path for upward forces is different from that for downward forces. The structurally attached systems are the only type that the load path is the same for both upward and downward paths. The resulting upward load forces caused by external wind forces are calculated depending on the arrangement of the PV systems.

Wind resistance for low-sloped roofs can be analyzed by being dividing the roof area into three zones: "field (or interior), perimeter (or end) and corner zones" (Kirby 2011). Corner zones experience the greatest wind loads, the interior the least, and the perimeter zone experiences wind loads between the two extremes. The conical vortices and an accelerated flow region associated with oblique or cornering winds are shown in Fig. 3.1.

These wind loads may have the power to move rooftop PV equipment, "especially when not structurally attached to the building. The possible consequences of racks falling off the rooftops to racks sliding across the membrane could lead to injured persons and damaged property" (Kirby 2011).

The effects of these wind forces have been studied to develop the best practices guides and code requirements for installing, testing, and maintaining PV systems. Preventative measures can ensure these loads do not surpass the structural capacity of the building. A challenge arises in predicting the maximum load a PV panel may create on a roofing structure in such an event. The Solar American Board for Codes

© Fire Protection Research Foundation 2015
R. Wills et al., *Best Practices for Commercial Roof-Mounted Photovoltaic System Installation*, SpringerBriefs in Fire, DOI 10.1007/978-1-4939-2883-5_3

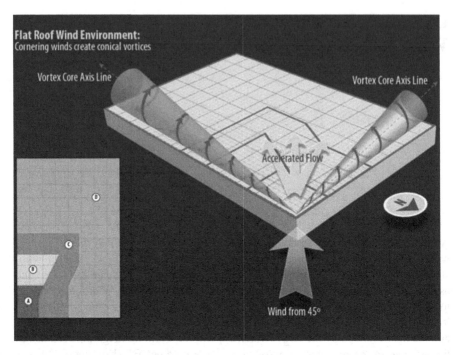

Fig. 3.1 Corner wind effects (Cermak Peterka Petersen)

and Standards addresses this issue in the "Wind load Calculations for PV Arrays (2010)" report prepared by Stephen Barkaszi, P.E. and Colleen O'Brian, P.E. This report mentions, "*ASCE Standard 7-05* is the standard for evaluating wind forces on structures." The *ASCE Standard 7-05* "provides requirements for general structural design and includes means for determining dead, live, soil, flood, wind, snow, rain, atmospheric ice, and earthquake loads, and their combinations that are suitable for inclusion in building codes and other documents" (Thomson Routers 2013).

The equations found in ASCE 7-05 are based on Eq. (3.1) based on fundamental fluid dynamics.

$$\text{Wind pressure} = \frac{1}{2}\rho * v^2 * C \qquad (3.1)$$

where

ρ density of air
v wind velocity
C dimensionless pressure coefficient measured for a side of a specific object.

"The most challenging part to estimating wind loads on any structure is finding out which of the many pressure coefficients in *ASCE 7-05* should be used. The pressure coefficient depends on many factors, including the shape of the structure and the tributary area of the structural component being analyzed" (O'Brien and

Banks 2012). The tributary area is the entire area that is being subjected to the loading through wind uplift.

Although this standard can be helpful, it does not provide information on how to assess wind loads on PV system installations.

"While some data are provided in the *ASCE 7* standard related to 'rooftop equipment,' these were developed for equipment with a prismatic shape, such as chimneys and HVAC units, with no gaps between the equipment and the roof. These data are not applicable to roof-mounted PV. Designers are left with nothing to do but guess which tables and figures for example, which building shapes—in the building codes best apply to PV systems. Many of the choices designers must make depend upon the type of building classification" (O'Brien and Banks 2012). This hybrid approach can lead to unconservative results especially when considering corner roof zones (SEAOC Report PV2-2012).

In response to the lack of methods to determine wind loads, the Structural Engineers Association of California (SEAOC) developed a document "Wind Design for Low-Profile Solar Photovoltaic Arrays on Flat Roofs" (SEAOC Report PV2-2012). The SEAOC PV Committee consisted of building officials, industry members, wind tunnel research experts, and members of other national code committees and the SEAOC seismology committee (Ward 2013). The paper describes the wind flow characteristics on rooftop PV systems. The same design methodology contained in ASCE 7-05 is incorporated into the SEAOC document. The requirements for wind tunnel studies, effective wind area computation, and the wind loads on the roof itself are also addressed. This document incorporates the lessons learned on wind tunnel testing in the study by Kopp (2012).

A wind tunnel study performed at the University of Western Ontario shows that there are two main mechanisms that cause aerodynamic loading: turbulence generated by the systems and pressure equalization (Kopp et al. 2013). This study highlights the difference between a PV system mounted to a roof versus one mounted to the ground. The very presence of the building changes the aerodynamic load. Basically, a very important aspect of roof-mounted PV panels is the complexity that the panel adds to the building aerodynamics. The gaps between the rooftop and the PV systems will "allow pressurization below the surface of the PV modules independent of pressure in the building interior" (O'Brien and Barkaszi 2010). This pressure can cause significant loading to the roof, which can be damaging.

This lack of guidance creates obstacles for the PV industry resulting in problems that include frustrated installers, dissatisfied customers, and wind related structural failures. "In addition, uncertainty about what constitutes a safe and secure installation for a given wind load can slow or stop the approval process for PV installations and complicated the training of code officials" (PV Racking and Attachment Criteria for Effective Low Slope Metal Panel Roof System Integration 2013).

One of the first methods that tried to incorporate elements of PV systems using the fundamentals in ASCE 7-05 is the DNV Wind Load Calculator for Sloped PV Arrays on Flat Roofs. "The DNV Wind Load Calculator uses an alternative method of calculating wind loads that was developed by determining which parts of *ASCE 7-05* match available wind tunnel data" (O'Brien and Banks 2012). Although this may seem like an easy immediate solution it is explained that it may not be the most scientific device. "The DNV Wind Load Calculator's method is completely ad hoc; there is little justification for it on the basis of physics or the intent of the code. It just fits the data" (O'Brien and Banks 2012).

This calculation can be used for ballast-only, module, and structurally attached systems. The user can enter the data used for the ASCE 7-05 calculation such as mean roof height, basic wind speed, directionality factor, importance factor, topographic factor, velocity pressure exposure coefficient, and effective wind area. The next calculation uses effective wind area or tributary area, PV tilt, row spacing, height above the roof, sheltering from wind due to nearby objects such as nearby subarrays or parapet walls, and location of the array being considered. Although this free tool is a step in the right direction, it does not allow the user to understand the basis of the calculations or how to alter them for a unique situation.

The wind study performed at the Western University used the third design method described in ASCE 7-05. This alternative method permits the use of wind tunnel testing as the basis for design. There are many requirements for wind tunnel testing in order to avoid biased results. For example, the atmospheric boundary layer has to be modeled to account for the variation of wind speed with regard to height, and the projected area of the modeled building and surroundings must be less than eight percent of the test section cross sectional area for the wind study (Kopp et al. 2013). The following three tests were conducted:

- Aeroelastic fly away—wind speed is increased until model moves
- Force balance—Strain gauges or load cells measure reaction forces at connections to the roof
- Pressure tap—A grid of pressure taps measures wind pressure on each surface (module faces)

Further details of these methods including the type of results obtained per tunnel run, amount of data collected, and key challenges are shown in Table 3.1.

"Wind tunnel testing can provide an appropriate basis for design of rooftop solar arrays per the code if the testing is done properly and the results of these test are properly applied" (Kopp et al. 2013). However, conducting this method can be very complex and expensive. Few facilities can meet the minimum requirements of this method. "Because the cost, time and effort required to perform this type of testing for each specific PV project would be prohibitive, the challenge is to develop a set of test data that can be used to provide design loads for a wide variety of different buildings, sites and array shapes. This type of generalization is possible with the appropriate test program, but is a complex and challenging undertaking" (Kopp et al. 2013).

The SEAOC report also incorporates the guidelines set by the California Department of Forestry and Fire Protection in their 2008 guideline. In this document Eq. (3.1) is edited to incorporate more factors such as the solar panel height above the roof and the low edge and raised edge, chord length of the solar panel, width of the overall building, parapet height factor, array edge increase factors,

Table 3.1 Wind tunnel testing methods and applicability to rooftop PV arrays (Kopp 2011)

Type of model	Aeroelastic fly-away	Force-balance	Pressure tap
Description	Wind speed increased until model moves (e.g., slides or lifts)	Strain gauges or load cells measure reaction forces at connections to the roof	A grid of pressure taps measures wind pressure on each surface (mod. faces)
Results obtained	Wind speed at which model overcomes friction and/or gravitational force	Reaction force coefficients for a range of wind speeds	Local and overall design pressure coefficients for a range of wind speeds and module areas
Measurement of local pressure peaks	"Local" defined as area of model that can move independently. Therefore not typically measured at the module level or smaller	"Local" defined as area of the model that can be shown to contribute to load on a given load cell. Therefore not typically measured at the module level or smaller	"Local" defined as area attributable to a given pressure tap. Measurement at module and sub-module level is therefore possible
Measures downward forces on array or roof?	No	Yes	Yes
Data points gathered per wind tunnel run	1	$\sim 1 \times 10^5$ to 1×10^6	$\sim 1 \times 10^7$
Key challenges/ drawbacks	• Large numbers of experiments required to generate reliable data set • Fidelity of scaled model (geometry, weight, stiffness) is critical to accuracy of results and difficult to attain • Cannot be easily applied to arrays of different sizes or shapes	• Large data sets to manipulate and analyze • Instrumentation is difficult at small model scales • Extensive computational and analytic capability required to apply results to arrays of different sizes, shapes and ballast configurations	• Large data sets to manipulate and analyze • Instrumentation required to accurately measure pressures on all appropriate surfaces is challenging • Extensive computational and analytic capability required to apply results to arrays of different sizes, shapes and ballast configurations
Feasibility	Scale of testing required for reliable and applicable results can be cost prohibitive. Sufficiently accurate model may be beyond state of the art depending on scale and complexity of structure	Currently difficult to instrument at scale needed	Is the most common method of testing to determine design requirements for wind loads on low-rise buildings and building components

Design Wind Loads	$h \le 60$ ft, or all heights if $h < W_S$
Figure 29.9-1(cont.) Roof Mounted Solar Panel Arrays	Roofs $\theta \le 7°$
Enclosed, Partially Enclosed Buildings	

Notes:

1. (GC_{rn}) acts towards and away from the panels top surface.
2. There shall be a minimum air gap around the perimeter of each solar module of 0.5 inches or between rows of panels of 1 inch to allow pressure equalization above and below panels.
3. Alternatively, for $\omega = 0°$, $h_1 \le 10''$, and air gap per note 2, use components and cladding procedure per ASCE 7-10 30.4 (ASCE 7-05 6.5.12.4).
4. Array should not be closer than $2(h_2 - h_{pt})$ or 4 feet, whichever is greater, from roof edge.
5. Roof structure area covered by solar array need not to be designed for simultaneous application of solar array wind loads and roof components and cladding wind loads. As a separate load case, roof structure shall also be designed for full roof components and cladding wind loads assuming PV panels are not present.
6. Notation:

A:	Effective wind area for structural element being designed, in ft.[2]
A_n:	Normalized wind area, equal to $\left(\dfrac{1000}{(max\,(a_{pv},\,15ft\,))^2}\right)A$
a_{pv}:	$0.5\sqrt{hW_L}$, but need not exceed h, in ft.
d_x:	Horizontal distance measured orthogonal to the panel edges in the north (d_N), south (d_S), east (d_E) and west (d_W) direction, from panel being evaluated to adjacent panel or building edge, whichever is closer, ignoring any rooftop equipment, in ft. For panels in a row, d_E and d_W are measured from the end of the row in their respective direction. E_E and E_W apply only to the panels within 5 ft of each end of the row on their respective side, and panels greater than 5 ft from both ends of their row shall have d_E and d_W =0.
E:	Array edge factor calculated for each panel area in each principle direction at a time, equal to maximum of E_N, E_S, E_E, E_W. If panel area being evaluated is located in zone 2 or 3 and d_N measured to building edge ignoring all other panels is greater than 3 a_{pv}, then E_N for that panel area need not exceed 1.5. If panel area being evaluated is located in zone 2 or 3 and d_S, d_E, or d_W measured to building edge ignoring all other panels is greater than 3 a_{pv}, then E_S, E_E, or E_W for that panel area in only that respective direction need not exceed 1.0.
(GC_{rn}):	Net pressure coefficient, equal to $\gamma_p E[(GC_{rn})_{nom}(\gamma_c)]$
$(GC_{rn})_{nom}$:	Nominal net pressure coefficient.
h :	Mean roof height above ground, except for monoslope roofs use maximum roof height, in ft.
h_1:	Solar panel height above roof at low edge, in ft.
h_2:	Solar panel height above roof at raised edge, in ft.
h_c:	Characteristic height, equal to $min(h_1,\,1ft)+l_p sin(\omega)$, except when evaluating E toward a building edge unobstructed by panels, then $h_c = 0.1 a_{pv}$ for that panel in that direction, in ft.
h_{pt} :	Mean parapet height above adjacent roof surface, in ft.
l_p:	Chord length of solar panel, in ft.
W_L :	Width of overall building on longest side, in ft.
W_S:	Width of overall building on shortest side, in ft.
γ_c:	Panel chord length factor, equal to 1.0 for $\omega < 5°$, equal to $0.6 + 0.06 l_p$ for $\omega > 15°$ but shall not be less than 0.8. For $5° < \omega < 15°$, apply γ_c only to $15°- 35°$ $(GC_{rn})_{nom}$ figure values and prior to interpolating.
γ_p:	Parapet height factor, equal to 1.0 for $h_{pt} \le 4$ ft, equal to the smaller of $0.25 h_{pt}$ and 1.3 for $h_{pt} > 4$ ft.
θ:	Angle of plane of roof from horizontal, in degrees.
ω:	Angle of plane of panel to roof, in degrees.

Fig. 3.2 Wind load calculations (SEAOC Report PV2-2012)

among others. This advanced system for calculating wind loads is displayed in Figs. 3.2 and 3.3 (SEAOC Report PV2-2012).

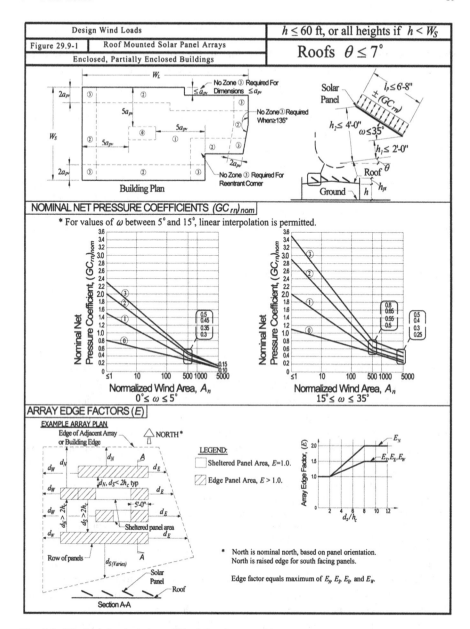

Fig. 3.3 Wind load calculations (SEAOC Report PV2-2012)

The following are the steps to calculate wind loads based on the Figs. 3.2 and 3.3.

Step 1:   Confirm applicability of the figure to the solar installation and building
Step 2:   For panels parallel to the roof surface and height above roof less than
           10 inches, the alternate procedure using components and cladding
           procedure
Step 3:   Confirm that layout provides minimum distance from roof edge
Step 4:   Compute a_{pv} for the building
Step 5:   Determine roof zones with respect to solar array layout
Step 6:   Determine effective wind area for each element being evaluated
Step 7:   Compute $(GC_{rn})_{nom}$ from applicable chart
Step 8:   If using 15 to 35 degree chart, apply chord length adjustment factor, γ_c
Step 9:   For tilt angles between 5 and 15 degrees, interpolate for tilt angle
Step 10:   Apply edge factors to edge rows, sides, and all rows where space
           between rows exceeds $2*h_c$
Step 11:   Apply parapet height factor, γ_p
Step 12:   Calculate GC_{rn} using γ_p, E, and $\gamma_c*(GC_{rn})_{nom}$
Step 13:   Go to ASCE 7-10 Table 29.1-1 to complete calculation (SEAOC Report
           PV2-2012).

This is a more accurate way of determining external wind forces than is described
in the wind loading Figs. 3.2 and 3.3. A report by Banks (2013) discusses the
difference between the SEAOC Report PV2-2012 and previous methods of calculating wind forces. "Several aspects of these procedures can be considered significant departures from how wind loads are calculated for the roof itself, including the
size of the edge and corner zones, the influence of parapets, and the use of an
effective tributary area that is normalized by the size of the building" (Banks 2013).

> The development of a wind loading figure for roof mounted solar photovoltaic arrays that
> corresponds to the prescriptive method in ASCE 7-05 is challenging due to the complexities
> of the wind flow characteristics on a roof and the numerous possible array layouts, con-
> figurations, and geometry. The goal is to make a simple, easy-to-use figure that fits most
> low-profile solar photovoltaic installations within the range of sizes and configurations most
> commonly used (SEAOC Report PV2-2012).

The SEAOC method is the most accurate and cost effective way available to
calculate wind forces on PV systems. There is still progress to be made with this
system to make it more specific for each situation. A future goal is to have this
report be adopted by the International Code Council. Wind design of roof-mounted
PV systems will be addressed in ASCE 7 in 2016, with possible adoption by the
International Building Code (IBC) in 2018.

Although this resource is the most effective way to calculate wind forces, it still
has its limitations. Structurally attached and ballast-only PV systems are the only
attachment method that has a determined way of predicting wind forces in every
direction. Many PV systems are not structurally attached and need an alternate
method to predict these wind forces. The SEAOC Report PV2-2012 does not have

any guidelines on how to calculate effective wind area. A report by Maffei (2014) "Wind Design Practice and Recommendations for Solar Arrays on Low-Slope Roofs" has described a process for determining effective wind areas.

For each gust load case, determine the appropriate gust load area A_{ti} and effective wind area, as follows:

(1) Calculate the tributary area for the element of interest. Per the definition of effective wind area in ASCE 7-10 (ASCE 2010), the effective wind area used to determine GC_p is the same as the tributary area except that the width of the effective wind area need not be taken less than one-third its length. The gust pressure is then applied on the actual area tributary to the element (Maffei 2014).

GC_p noted in the ASCE 7-10 standard is the pressure coefficient. This step is only a portion of the process but this report along with another report by Schellenberg (2013) describes how to determine the effective wind areas. Through testing and modeling these reports describe a more advanced way to determine wind loading.

One of the principle authors of the SEAOC Report PV2-2012, Kopp, notes that deflectors/shrouds are not addressed within SEAOC Report PV2-2012. Shrouds affect the wind loads being applied to the PV system; "It is not wise to assume that loads from one racking system apply to another unless the geometries are quite similar. This is especially true for racking systems that feature deflectors or shrouds because small changes in design can markedly improve loads in one region of the roof and yet worsen loads elsewhere" (Kopp 2012).

Chapter 4
Hail

Hail is another hazard that is introduced by weather conditions. Hail has the potential to damage PV systems. The impact of hail's impulsive force acting on the PV system can cause cracking of PV systems. This can damage the PV system's ability to convert energy as well as introduce exposed electrical hazards.

The International Electrical Code (IEC) addresses how to conduct an impact test for PV systems. This impact test demonstrates that a PV system can withstand an expected hailstorm.

The respective standard is based on the design of the modules; IEC 61215 is used for crystalline silicon, IEC 61646 for thin film, and IEC 62108 is used for concentric PV Modules. Each type of module introduces different hazards because of the various materials, designs, and process flaws that may lead to premature field failures. In order to represent long term effects of outdoor conditions, accelerated stressed tests are used (Wohlgemuth 2012). These accelerated stress tests for hail use a series of impacts of ice balls at various speeds. Modules that have passed the qualifications test are more likely to survive in the field because they have met specific requirements that allow them to withstand hazards posed in the field (Wohlgemuth 2012).

A report by Regan Arnt (2010), Basic Understanding of IEC Standard Testing For Photovoltaic Panels, describes the required Hail test for the IEC standards (Table 4.1).

Hail impact: is a mechanical test. To verify that the module is capable of withstanding the impact of hailstones which are at a temperature of ~ -4 °C. The test equipment is a unique launcher capable of propelling various weights of ice balls at the specified velocities so as to hit the module at 11 specified impact locations +/− 10 mm distance variation.

The time between the removal of the ice ball from the cold storage container and impact on the module shall not exceed 60 s.

It is quite common practice to use 25 mm/7.53 g ice balls.

Again, after the test one should check if there are any major defects caused by the hailstones, and also P_{max} (for IEC 61215 only) and insulation resistance are checked.

Laboratory statistics show very low failure rates for this test.

© Fire Protection Research Foundation 2015
R. Wills et al., *Best Practices for Commercial Roof-Mounted Photovoltaic System Installation*, SpringerBriefs in Fire, DOI 10.1007/978-1-4939-2883-5_4

Table 4.1 Ice ball masses and test velocities (Arnt 2010)

Diameter (mm)	Mass (g)	Test velocity (m s^{-1})	Diameter (mm)	Mass (g)	Test velocity (m s^{-1})
12.5	0.94	16.0	45	43.9	30.7
15	1.63	17.8	55	80.2	33.9
25	7.53	23.0	65	132.0	36.7
35	20.7	27.2	75	203	39.5

This test demonstrates that a PV system can withstand the impact forces from hail. Many PV systems are tested in accordance with this standard.

American Standards for Testing and Maintenance ASTM E1038 is another testing standard that is used to determine the resistance of PV systems to hail. This standard uses propelled ice balls to simulate hailstones. The effects of impact may be either physical or electrical degradation of the module. The testing standard specifies the proper method for mounting the test specimen, conducting the impact test, and reporting the effects. The mounting method tested depends on the arrangement that will be used in real life scenarios. Different impact locations are determined based on vulnerable areas on the array. The size and weight of the ice balls are also specified. The velocities of the ice balls are meant to be comparable to speeds that real hailstones could hit a PV system during a storm. The ASTM E1038 standard does not establish pass or fail levels but instead provides a procedure for determining the ability of photovoltaic modules to withstand impact forces of falling hail.

UL 1703 "Flat-Plate Photovoltaic Modules and Panels" has a Section 30 "Impact Test" that is used on many panels before they are installed. Within this testing guideline it states that there may be no particles larger than 1 in^2. (6.5 cm^2) released from their normal mounting position. A 2 in. steel ball is used to represent a hailstone falling onto the panel. As in ASTM E1038 the mounting of the PV system is to be representative of its intended use. Other testing procedures are described such as the distance that the ball must fall from, the location of the impact, among others, that ensure that the test is representative of an actual hailstone striking the panel.

Hail hazards are also addressed by a set of FM Global's Approval Standards. There are two types of modules that FM Global refers to: rigid modules and flexible modules. "Rigid modules (or crystalline silicon) modules are currently the most common form of solar energy system, and typically require a metal rack system for roof or ground mounting. Flexible PV (thin film) modules secured to roofing assemblies currently represent a small, but rapidly growing segment of the overall solar energy market" (FM Global 2011). The Approval Standard for the Rigid PV systems is FM 4478, and the Approval Standard for the Flexible PV systems is FM 4476.

The following section on hail resistance testing is from the FM Approval Standard 4478, applicable to rigid photovoltaic modules (2012):

4.5 Hail Damage Resistance Test

4.5.1 Testing for hail damage resistance shall be in accordance with Test Procedure, Test Method for Determining the Susceptibility to Hail Damage of Photovoltaic Modules, FM Approvals, LLC. The minimum rating required for FM Approval is Class 2.

4.5.1.1 Condition of Acceptance for Hail Damage Resistance

4.5.1.2 After completion of the impacting testing, the photovoltaic module shall show no signs of cracking or slipping or misaligned external surfaces, or rupture when examined closely under 10X magnification.

FM Approval Standard 4478 outlines how to test the system and refers to another Approval Standard, FM 4473—Specification Test Standard for Impact Resistance Testing of Rigid Roofing materials by Impacting with Freezer Ice Balls. "This testing standard provides a procedure for determining the impact resistance performance of new prepared rigid roofing materials" (FM Global 2005). "Ice balls are used in this test method to simulate hailstones. Hailstones are variable in properties such as shape, density, and frangibility. These properties affect factors such as duration and magnitude of the impulsive forces acting on the roof area which the impulse is distributed" (FM Global 2005). Ice balls are typically more dense than hailstones so ice balls present a worst case hailstone.

The testing parameters control the size and mass of the ice balls, the size of the test specimen, the temperature of the specimen, the angle of the specimen, perpendicular trajectory of the ice ball to the specimen, and the speed of the ice ball, which controls the impact kinetic energy. Any damage done to the panel is assessed and reported with the impact locations marked (FM Global 2005). Although this is a thorough test standard, there is little testing that is actually being done on PV systems.

In 2012, one test was performed R.B. Uselton at the Applied Research Group for Lennox Industries. These tests were not done following the FM Approval Standard 4473. The minimum size test balls that were supposed to be used was $1\frac{1}{4}$ in. ice balls; this test used 1 in. and $1\frac{1}{3}$ in. ice balls. The size of the ball has a significant affect on the impact energy transferred to the panel. Only eleven ice balls were used for testing, this is far less than what a PV system would have to withstand during a hailstorm. There are regions in the world that are prone to severe hail and could experience larger hailstones than the hailstones that are used in any of these tests.

Hail is an example of the effect of weathering on PV systems. Long-term degradation may also occur as a result of the effects of temperature cycling, ultraviolet exposure and corrosion. Another section of this report discusses further the overall effect of long-term weather related hazards. Speaking generally about qualification standards such as IEC61215, John Wohlgemuth points out, "Although these testing standards are helpful in creating PV systems that can withstand early infant mortality rate, there are limitations to these tests. These tests do not identify and quantify wear-out mechanisms. They do not address failure mechanisms for all climates and system integrations. Finally, these tests are not meant to quantify the lifetime for the intended application/climate" (Wohlgemuth 2012).

Chapter 5
Snow

Snow can affect both the energy performance of a PV system as well as the structural integrity of the PV system. A report by Ross (1995); "Snow and Ice Accumulation on Photovoltaic Arrays: An Assessment of the TN Conseil Passive Melting Technology," assesses the ability of a PV system to melt the snow off the panels. "When the surface of a snow- or ice-covered PV panel attains a temperature higher than that of the snow or ice on its surface, heat is transferred to the snow or ice." "The panel tilt angle not only affects snow and ice accumulation on a PV panel, but, through gravity, it determines the force motivating the snow or ice to slide off the panel" (Ross 1995). Although this report discusses how snow can be removed from the panels, it does not discuss the structural loading impacts of the accumulated snow. Many other reports cover the performance losses from snow but again do not cover the structural impacts.

An assessment of possible snowdrift build-up should be considered in the construction and installment of PV systems in areas where snow is expected (Racking and Attachment Criteria for effective Membrane Roof System integration, July 2012). In order to provide a complete assessment, snow loads could be calculated in a similar manner as wind loads. However, documentation of this type of analysis is limited. Snow loads are addressed in the IEC 61215. An accelerated stress test using static mechanical loading for PV systems represents loads that snow buildup would introduce. On a related issue, the IEC 61215 also has a salt spray test that is used for determining corrosion due to salt used for removal of snow and ice.

© Fire Protection Research Foundation 2015
R. Wills et al., *Best Practices for Commercial Roof-Mounted Photovoltaic System Installation*, SpringerBriefs in Fire, DOI 10.1007/978-1-4939-2883-5_5

Chapter 6
Debris Accumulation

Debris accumulation is another major hazard applicable to both roofs and PV systems. Partial shading is a problem that can arise from dirt buildup on module surface. Partial shading can decrease the effectiveness of the PV panels, which may dissuade consumers from accepting the new energy source. Having to periodically clean roof-mounted PV panels to eliminate cell shading may subject workers to increased incidence of fall and shock injuries.

Debris build up can be a result of undrained water floating on top of roofs. "Construction tolerances and roof slopes are often so low that the mere presence of a roof seam may limit drainage. This is especially true when seams cross the drainage flow" (Kirby 2012). Not only does debris affect the efficiency of the panel, debris can also quickly turn into a fire hazard. Underwriters Laboratory (UL) created a study to determine how well screens would prevent ignition of debris accumulation between mounted PV systems, also providing a significant amount of information on how fire spreads between PV panels and roof tops.

The main point from the study was that without any debris, a class A rated roof with a 5″ mounted PV panel resisted ignition when exposed to hot embers; therefore making the assumption that the roof rating remains true even with a PV panel mounted on it. Yet, when debris accumulated between the PV panels and the roof, even when protected by 1/8″ and 1/16″ screens, and it was then exposed to hot embers, the debris ignited spreading the fire across the roof. Even though the roof was a Class A rated roof, the containment and spread of the fire between the panels and the roof caused damage to both the roof and the PV system. Though the screens proved to not prevent debris ignition, screens could be used to prevent the accumulation in the first place (Backstrom 2012).

UL researched the effects of the screens on the PV systems operating temperatures. This research discovered that the addition of screens tended to increase the operating temperature of the PV system, and though further research has not yet been done, the knowledge of how components react in greater operating temperatures raises the concern that adding screens could lead to earlier component failure. Earlier component failure is discussed further in the Fire Hazards section.

© Fire Protection Research Foundation 2015
R. Wills et al., *Best Practices for Commercial Roof-Mounted Photovoltaic System Installation*, SpringerBriefs in Fire, DOI 10.1007/978-1-4939-2883-5_6

The accumulation of debris such as dirt, snow, and hail is a hazard that is being investigated due to its potential to damage the roofing structure in a PV system. Ideally, adhered PV products are installed running parallel to the slope of the roof. This helps reduce dirt accumulation and standing water. This also minimizes the cell shading that compromises PV performance (Kirby 2011).

"PV Racking and Attachment Criteria for Effective Low-Slope Roof System Integration" by The Center PV Taskforce provides recommendations for five fundamental principles of effective roof system integration concerning external forces, system integration, roof drainage, roof and PV system maintenance, and roof safety. The roof drainage section describes solutions for preventing and dealing with sitting roof water and debris accumulation. For example, limiting horizontal elements, providing accessible roof drains for periodic maintenance, and providing walkway areas for roof inspection and maintenance all help prevent debris build up.

Chapter 7
Seismic

SEAOC developed a draft document that addresses the seismic hazards associated with rooftop PV systems; "Structural Seismic Requirements and Commentary for Rooftop Solar Photovoltaic Arrays" (SEAOC Report PV1-2012). The seismic requirements document is applied to the three types of PV systems; unattached (ballast-only), attached roof-bearing systems, and fully framed systems.

If a PV system is added to an existing structure the seismic force resisting system of the building should be checked per the requirements of Chapter 34 of the IBC 2009. "Per Sections 3403.4 and 3404.4 of IBC 2009, if the added mass of the PV system does not increase the seismic mass tributary to any lateral-force-resisting structural element by more than 10 %, the seismic-force-resisting system of the building is permitted to remain unaltered" (SEAOC Report PV1-2012).

For each of the three attachment methods there are separate requirements:

Fully framed systems:

PV support systems that are attached to the roof structure shall be designed to resist the lateral seismic force F_p specified in ASCE 7-05 Chapter 13 (SEAOC Report PV1-2012).

Attached roof-bearing systems:

For attached roof-bearing systems, friction not to exceed $(0.9\ \mu_s - 0.2S_{DS})W_{pf}$, is permitted to resist the lateral force F_p where W_{pf} is the component weight providing normal force at the roof bearing locations, and μ_s is the coefficient of friction at the bearing interface. The resistance from friction is permitted to contribute in combination with the design lateral strength of attachments to resist F_p (SEAOC Report PV1-2012).

Unattached (ballast-only):

Unattached (ballast-only) systems are permitted when all of the following conditions are met:

- The maximum roof slope at the location of the array is less than or equal to 7 degrees (12.3 percent).
- The height above the roof surface to the center of mass of the solar array is less than the smaller of 36 inches and half the least plan dimension of the supporting base of the array.

© Fire Protection Research Foundation 2015
R. Wills et al., *Best Practices for Commercial Roof-Mounted Photovoltaic System Installation*, SpringerBriefs in Fire, DOI 10.1007/978-1-4939-2883-5_7

- The system is designed to accommodate the seismic displacement determined by one of the following procedures:

    - Prescriptive design seismic displacement
    - Nonlinear response history analysis
    - Shake table testing (SEAOC Report PV1-2012).

For unattached (ballast-only) systems, the PV system shall be designed to accommodate seismic displacement. The design seismic displacement of the array relative to the roof, Δ_{MPV}, can be used to allow sliding based on determined minimum separations between separate solar arrays, fixed objects, and roof edges. The strength requirements of each separate array are based on the total weight of the array.

Each separate array shall be adequately interconnected as an integral unit such that for any vertical section through the array, the members and connections shall have design strength to resist a total horizontal force across the section, in both tension and compression, equal to $0.1W_1$
Where
W_1 = the total weight of the array, including ballast, on the side of the section that has smaller weight.
The horizontal force shall be applied to the array at the level of the roof surface. The force $0.1W_1$ shall be distributed in proportion to the weight that makes up W_1. The computation of strength across the section shall account for any eccentricity of forces (SEAOC Report PV1-2012).

These separation distances only apply to structural requirements; they do not account for required separation distances needed for fire fighting access or electrical requirements. "The minimum clearance around solar arrays shall be the larger of the seismic separation defined herein and minimum separation clearances required for fire fighting access" (SEAOC Report PV1-2012).

Prescriptive design seismic displacement, Δ_{MPV}, can be used if the following requirements are met:

- I_p per ASCE 7-05 Chapter 13 is equal to 1.0 for the solar array and for all rooftop equipment adjacent to the solar array.
- The maximum roof slope at the location of the array is less than or equal to 3 degrees (5.24 percent).
- The manufacturer provides friction test results, per the requirements herein, which establish a coefficient of friction between the PV support system and the roof surface of not less than 0.4. For Seismic Design Categories A, B, or C, friction test results need not be provided if the roof surface consists of mineral-surfaced cap sheet, single-ply membrane, or sprayed foam membrane, and is not gravel, wood, or metal.

Δ_{MPV} shall be taken as follows:
Seismic Design Δ_{MPV} Category:

- A, B, C – 6 inches
- D, E, F – $[(SDS- 0.4)^2] * 60$ inches, but not less than 6 inches (SEAOC Report PV1-2012).

Any electrical connections must be able to accommodate these seismic displacements. There are different types of testing that can be done on the PV system and the roof that will represent the forces that would occur on the system if a seismic event were to occur. One type of testing is called friction testing, the following is an excerpt from (SEAOC Report PV1-2012) describing friction testing;

"The coefficient of friction used in these requirements shall be determined by experimental testing of the interface between the PV support system and the roofing surface it bears on. Friction tests shall be carried out for the general type of roof bearing surface used for the project under the expected worst-case conditions, such as wet conditions versus dry conditions. The tests shall conform to applicable requirements of ASTM G115, including the report format of Section 11. An independent testing agency shall perform or validate the friction tests and provide a report with the results.

The friction tests shall be conducted using a sled that realistically represents, at full scale, the PV panel support system, including materials of the friction interface and the flexibility of the support system under lateral sliding. The normal force on the friction surface shall be representative of that in typical installations. Lateral force shall be applied to the sled at the approximate location of the array mass, using displacement controlled loading that adequately captures variations in resistive force." "If stick-slip behavior is observed, the velocity shall be adjusted to minimize this behavior. Continuous electronic recording shall be used to measure the lateral resistance. A minimum of three tests shall be conducted, with each test moving the sled a minimum of three inches under continuous movement. The force used to calculate the friction coefficient shall be the average force measured while the sled is under continuous movement. The friction tests shall be carried out for the general type of roofing used for the project" (SEAOC Report PV1-2012).

Other types of testing are nonlinear response history analysis and shake table testing, below is an excerpt from SEAOC Report PV1-2012 describing requirements for these types of testing.

The design seismic displacement corresponding to the Design Basis Earthquake shall be determined by nonlinear response history analysis or shake table testing using input motions consistent with ASCE 7-05 Chapter 13 design forces for non-structural components on a roof.

The analysis or test shall use a suite of not less than three appropriate roof motions, spectrally matched to broad band design spectra per AC 156 Section 6.5.1. Each roof motion shall have a total duration of at least 30 seconds and shall contain at least 20 seconds of strong shaking per AC 156 Section 6.5.2. The spectrum shall vary linearly with component period T in the increasing portion of the acceleration-sensitive region, and shall be proportional to $1/T$ in the velocity-sensitive region. A three-dimensional analysis model or experiment shall be used, and the roof motions shall include two horizontal components and one vertical component.

The analysis model or experimental test shall account for friction between the system and the roof surface, and the slope of the roof. The friction coefficient used in analysis shall be based on testing per the requirements herein.

If at least seven roof motions are used, the design seismic displacement is permitted to be taken as 1.1 times the *average* of the peak displacement values (in any direction) from the analyses or tests. If fewer than seven roof motions are used, the design seismic displacement shall be taken as 1.1 times the *maximum* of the peak displacement values from the analyses or tests. Roof motions shall have a minimum duration per AC 156 consistent with the expected Design Basis Earthquake motions at the site.

Resulting values for Δ_{MPV} shall not be less than 50 % of the values specified in Section 6, unless lower values are validated by independent Peer Review (SEAOC Report PV1-2012).

Testing has been done by SunLink on ballast-only systems. Seismic tests on two separate roof systems were conducted and it was demonstrated that no damage was done to the roofing assembly and did not compromise the integrity of the roofing membrane.

The following were used as objectives for the testing;

Observe the interaction between the PV systems and the roof membranes and document any areas of concern.

Use test results to validate and calibrate computer analysis models predicting PV system movement during an earthquake.

Provide additional documentation to building code officials reviewing and permitting ballast only mounted systems.

Ultimately, influence the creation of new structural engineering standards permitting unattached PV array installations similar to those ASCE 7-05, Chapter 17 permitting friction isolation systems (SunLink 2012).

A summary of the actual process and methods used in the testing can be found in "SunLink Seismic Testing Roof Integrity Report". It should be noted that this report is not detailed enough to determine that the tests done were in fact rigorous.

Although these testing methods have been developed to best represent the same forces that will be exerted on to a PV system and its roof, there are still limitations to these testing methods. One limitation is representing the behavior of frozen PV elements on top of a frozen or frost-covered roof subject to seismic motions. A requirement for this type of hazard is described by SEAOC:

For solar arrays on buildings assigned to Seismic Design Category D, E, or F where rooftops are subject to significant potential for frost or ice that is likely to reduce friction between the solar array and the roof, the building official at their discretion may require increased minimum separation, further analysis, or attachment to the roof (SEAOC Report PV1-2012).

Although the above requirement says to provide additional analysis so that rooftops that are subject to significant potential to freezing can be protected, the following commentary is included to show that fulfilling this requirement may be difficult:

The PV Committee is not aware of any research specifically addressing (a) the potential for frost or freezing of this type, (b) the effect of frost on the friction behavior of various roof surfaces, or (c) the likelihood that such frost forms underneath or sufficiently adjacent to solar panel feet as to compromise displacement resistance. Section C10.2 of ASCE 7-10 describes some of the phenomena related to the formation of frost, freezing rain, and ice (SEAOC Report PV1-2012).

Since forces on the top of the building are very different than those exhibited at the base of the building, shake table testing motions are scaled to represent that. The following SEAOC commentary excerpt describes this limitation and the associated hazards with it:

Nonstructural components on elevated floors or roofs of buildings experience earthquake shaking that is different from the corresponding ground-level shaking. Roof-level shaking is filtered through the building so it tends to cause greater spectral acceleration at the natural period(s) of vibration of the building and smaller accelerations at other periods. The target spectra defined in AC 156 are broadband spectra, meaning that they envelope potential peaks in spectral acceleration over a broad range of periods of vibration, representing a range of different buildings where nonstructural components could be located.

In lieu of spectrally matching (frequency scaling) motions to a broadband roof spectrum, it may also be acceptable to apply appropriately scaled Design Basis Earthquake ground motions to the base of a building analysis model that includes the model of the solar array on the roof. In such a case the properties of the building analysis model shall be appropriately bracketed to cover a range of possible building dynamic properties and behavior. Because friction resistance depends on normal force, vertical earthquake accelerations can also affect the horizontal movement of unattached components, so inclusion of a vertical component is required.

The factor of 1.1 used in defining the design seismic displacement is to account for the random uncertainty of response for a single given roof motion. This uncertainty is assumed to be larger for sticking/sliding response than it is for other types of non-linear response considered in structural engineering. The factor is chosen by judgment (SEAOC Report PV1-2012).

Chapter 8
Fire Hazards

The fire resistance ratings of roofs may be altered when a PV system is introduced to the roofing assembly. The electrically charged components, including the modules and wiring, pose a fire hazard. If not handled properly, electrical charges can be discharged as sparks, which could initiate a fire. The high operating temperatures of the PV systems also could affect the fire growth rate in the event of a fire. Two-sided vertical spread could occur in the gaps between the modules and the roof. The arrangement and spacing of the PV system can also aid or limit the horizontal fire spread. When structurally attaching the PV system to the roof, penetrations are made into the roof to make a place for the PV system supports. Even small holes can negatively affect the performance of a fire rated roof. Fire hazards associated with PV systems are discussed further in the following two subsections, flammability of components and ignition hazards.

8.1 Flammability of Components

The flammability characteristics of the components that make up a PV system depend on two factors: the materials used, and the arrangement of those materials. As described before, there are mounting devices that are used to align the panel at a specific angle towards the sun, which creates gaps between the PV panels and the roof. Different materials are used to support the PV modules themselves depending on the manufacturer and installer. Research has been conducted in the past few years to understand how these gaps affect flame spread and the fire resistance rating of roofs.

Early research results demonstrated that the fire class rating of the PV module alone (determined according to American National Standards Institute [ANSI]/UL 1703-2012) may not accurately predict the fire resistance properties of the system composed of PV array, mounting structure, and roof covering. Based on these results, stakeholders recognized the need to develop a new fire classification test for the PV module, mounting hardware, and roof assembly as one system. Over the last two years, the underlying

© Fire Protection Research Foundation 2015
R. Wills et al., *Best Practices for Commercial Roof-Mounted Photovoltaic System Installation*, SpringerBriefs in Fire, DOI 10.1007/978-1-4939-2883-5_8

principles of the new test have been refined and a new PV system flammability test regime was developed. In July 2013, the UL 1703 Standards Technical Panel (STP) adopted the new fire classification test into the standard (Rosenthal et al. 2013).

The gap between the PV system and the roof allows flames to transfer heat to both the underside of the PV system and to the outer layer of the roof. This narrow opening can vary depending on the angle and distance that the PV system is from the roof. Backstrom (2010) explains this gap further and how the size affects the temperature of the exposed surfaces:

For the parameters in this study, it was found that when the gap between the rack mounted PV module and the roof was reduced from 10 inches to 5 inches the measured surface temperatures increased. It was observed that both the 10 inch and 5 inch gap captured all of the flames, however the smaller gap also reduced the amount of entrained air into the fire plume thus elevating the temperature of exposed surfaces. When the gap size was reduced further to the value of 2.5 inch, the measured surface temperatures did not increase but rather lowered, as the gap was sufficiently decreased to capture only a portion of the flames. The influence of the setback of the PV module on the measured temperature and heat flux on the roof surface was highest when the PV module was in line with the leading edge (i.e., no setback distance). The measured temperatures and heat flux exposure lessened as the setback distance was extended.

This gap was analyzed, tested, and reported by Backstrom (2010) in "Effect of Rack Mounted Photovoltaic Modules on the Fire Classification Rating of Roofing Assemblies."

Results from Phase 1 study showed that installation of PV modules on roofs had an adverse affect on the Fire Class Rating of the roof assembly. For further confirmation, more experiments were conducted in Phase 2 with the PV module placed at the roof leading edge (0 in. setback distance) with a 5 in. gap between the PV module and roof surface. The results from PV modules on fire rated roof systems for the Spread of Flame tests are listed in Table 8.1. These results suggest that the presence of a PV module adversely affects the fire rating of a roof. If a roof is noncombustible, the flame spreads through the gap between the roof and the PV module in excess of 8 ft.

Some results of PV modules on fire rated roof systems for the Burning Brand Tests are described in Table 8.2. In the brand test involving Class A rated roof and Class C rated PV panel with the brand located on the roof, it was observed that the PV panel sagged and collapsed onto the roof allowing flames to vent vertically. This prevented the flame from penetrating the roof. However, though the results from two tests did meet the requirements for Class A, a third test did not. Since there was one case where the results were not in compliance with Class A requirements, clearly the random nature of fire growth and spread of the PV module affects the outcome of these particular tests.

Table 8.1 Influence of PV module during spread of flame test

Roof rating	PV rating	Flame spread
A	C	Greater than 8 ft.
A	A	Greater than 8 ft.
C	C	Greater than 8 ft.
Noncombustible	C	Greater than 8 ft.
Noncombustible	A	Greater than 8 ft.

Table 8.2 Brand test (Backstrom and Tabaddor 2010)

Roof rating	PV rating	Brand size/position	Fire performance result
A	C	Class A/PV	Compliant
A	C	Class A/Roof	2 Compliant/1 not compliant
C	C	Class C/Roof	Not compliant
A	A	Class A/Roof	Not compliant

As observed from the results of these tests, the arrangement of the PV systems as well as the brand of PV system affects the fire spread. These tests verified that the ratings of PV panels alone do not accurately describe the flammability character-istics of an entire PV system. Backstrom's further research indicated that a proper assessment of the flammability characteristics of a roof assembly with a PV system can only be obtained if a test is conducted with the particular PV system and roof assembly of interest.

As addressed previously, PV panels themselves do not represent the flamma-bility characteristics of an entire PV system that is attached to a roof. Code officials, members of the roofing industry and PV system industry have been concerned about the influence of PV systems on the fire resistance of a roof. FM 4478 Approval Standard also requires roof assembly and the PV to be tested as a system. Over the past few years this issue has been addressed through testing of stand-off mounted PV systems as described earlier. A test plan was developed to investigate the fire resistance properties of the stand-off configuration with funding provided by U.S. Department of Energy, UL, and the Solar America Board for Codes and Standards (Solar ABCs).

Sherwood (2013) describes the experiments that were conducted:

The new fire classification procedure requires the following tests be performed in order to derive a fire classification rating for the PV/roof system:

- spread of flame test on the top surface of module or panel,
- spread of flame test at roof and module or panel interface over representative steep or low sloped roof,
- burning brand test on module surface over representative steep sloped roof, and
- burning brand test between the module or panel and representative steep sloped roof.

Some of the major findings of the research program have been:

- For burning brand tests in which the brand is placed between the PV module and the roof surface, a Class B brand is the closest representation of actual materials likely to collect in this area.
- The critical flux values for ignition of low slope and steep slope roofing products and for crystalline silicon PV modules were found to be consistent for products in the same category (validating module 'typing' as a means to reduce testing requirements).
- The first to ignite (roof covering), second to ignite (PV) concept was demonstrated as a viable method for assessing the flammability performance of a system composed of PV, roof covering, and mounting hardware (Sherwood et al. 2013).

After these studies, in 2012, a new requirement in the IBC concerning the rating of PV panels was introduced. PV systems are now required to be the same fire classification rating of a roof as well as the PV system cannot affect the fire resistance rating of the roof. A white paper by Solar ABCs (2011) describes the impacts and difficulties associated with PV installations due to the changes of the 2012 international codes:

> The requirements of the 2012 IBC 1509.7.2 will need careful examination in their application. The language of this section states that the fire classification of PV systems must match the minimum fire classification of the roof assembly over which they are mounted as required in Section 1505. With any rooftop structure, the structure should not degrade the fire resistance properties of the roof, so as not to place the structure and its inhabitants at an unanticipated risk. However, straightforward implementation of this requirement is not possible.
>
> PV modules are a component of a rooftop mounted PV system and, although PV modules can receive a fire classification rating (in accordance with ANSI/UL 1703), there is presently no American National Standard Institute (ANSI) fire classification test or rating for a PV system. Similarly, there is no ANSI fire classification test for systems that include the PV array and the roof assembly. Thus, as currently written, Section 1509.7.2 refers to the fire classification rating of a system, and this exact approach is not yet available.
>
> In the absence of a PV system rating, it may seem appropriate to use the PV module fire classification rating in order to ensure the desired result, which is the preservation of the roof assembly's original fire classification. However, simply using the PV module fire classification rating may not provide the desired result in all cases.
>
> In 2008 and 2009, rigorous testing by UL and Solar ABCs revealed that the performance of a system—which includes PV modules on standoff mounted racks—exposed to fire or flame is not the same as that of a module alone. Currently, modules receive a fire classification rating based on testing the module alone, not as part of a PV system. The results of these tests show that actual performance of a rack-mounted PV system exposed to fire or flame is strongly dependent on the mounting geometry of the PV array and properties of the components that make up the specific module type. (A summary of this research is published in a Solar ABCs report available at: www.solarabcs.org/about/publications/reports/flammability-testing/index.html)
>
> As a result of this testing and in consideration of the current requirements of IBC Section 1509.7.2, Solar ABCs, UL, and an ANSI/UL 1703 Standards Technical Panel (STP) working group are actively developing a new test methodology. This is being done in close association with a working group composed of representatives from the PV industry, the roofing industry, standards development, the building and fire enforcement community, and government laboratory experts. The work product of this effort will be presented immediately to the full STP for UL 1703 for vetting and, ultimately, adoption. If adopted, this new test will be applicable to PV systems and will provide the valid, ANSI standard fire classification rating needed for compliance in the execution of the requirements of Section 1509.7.2.

The new requirements describe how to properly determine a fire rating for PV systems. While the standard was changed on October 25, 2013, it is important to note that that change is not effective until October 25, 2016 so it will not have any positive effect on this problem for several years. Although rated PV systems will soon be required, it is not understood how to apply this requirement. PV systems can be so unique that there is not a simple way to determine the fire resistance rating. The testing that has been done previously only covers a limited number of PV systems. Before 2013, there wasn't a standard that covered how to conduct fire resistance tests. UL 1703 was recently updated to include testing details that reflect those used in prior research efforts (Sherwood et al. 2013):

> The year 2013 marks a significant change for the fire classification rating approach for roof mounted stand-off photovoltaic (PV) modules and panels evaluated in accordance with American National Standards Institute/Underwriters Laboratories, Inc. (ANSI/UL) 1703, *Standard for Safety for Flat-Plate Photovoltaic Modules and Panels*. Prior to 2013, a PV module manufacturer could receive a fire classification rating based on tests of the module or panel alone. After the 2013 changes to ANSI/UL 1703, the fire classification rating approach takes into account the module or panel in combination with the mounting system and the roof covering products over which it is installed. The proposals that led to these changes were an outgrowth of research tests conducted and broad stakeholder forums held through a partnership between UL and the Solar America Board for Codes and Standards (Solar ABCs).

Having testing procedures available is an improvement for the PV industry. This will give designers a better ability to determine the fire rating of a PV system and how it affects the fire resistance rating of a roof. One concern about this approach is that most PV systems are unique and testing every single situation could become expensive (Sherwood et al. 2013) addresses this issue:

> The new fire classification rating tests in ANSI/UL 1703-2013 involve the combination of the module or panel, the mounting system, and the roof covering system. Because each of these three components has many products in the marketplace, testing every possible combination of the three components could mean thousands of required tests. This is not practical and could stifle market innovation. In response, a number of considerations and provisions were written into the new standard to reduce the number of required tests. In addition, Solar ABCs, UL, industry, and stakeholders continue to explore and validate industry-wide solutions that may satisfy the new, revised ANSI/UL 1703-2013 fire classification requirements in an effort to reduce the industry's testing burden.

Testing needed to be reasonable or else a drop in PV system installations could result. Although many issues of determining the fire rating of PV systems have been addressed, the experts still indicate that there is more that can be done.

> In July 2013, following stakeholder meetings and periods for public comment, the UL STP voted unanimously to approve the new fire classification test procedure. The STP acknowledged that further clarifications and refinements were still needed, including: defining additional PV module types in order to address new and old products not currently covered by the existing three types, and adding flexibility for the standard baseline roof types that meet the four-foot to six-foot fire performance criteria (Sherwood et al. 2013)

8.2 Ignition Hazards

The PV panels themselves can be combustible as well as the components that make up the entire system. Electrically charged components such as the panels and wiring also present an additional hazard when introduced to fire. Early component failure, especially in wiring, leading to an electrical fault, is often the source and the leading cause of PV system building fires. In the few PV system building fires that have been investigated, the causes for the fire have not been associated with the PV panels directly, but have actually been ground fault problems in the wiring. An interim report by Ball "Grounding Photovoltaic Modules: The Lay of the Land" summarizes the current state of codes and standards that apply to equipment grounding of PV systems. "The Solar America Board for Codes and Standards (Solar ABCs), commissioned this work with the intent of providing the PV industry with practical guidelines and procedures for module grounding. This initial "lay-of-the-land" report sets the stage for a final report that will draw on feedback from industry experts as well as ongoing research UL to develop guidelines and recommendations for changes to existing standards" (Ball 2011).

This is a concern because PV systems are typically installed on aluminum or steel frame structures that are electrically conductive and can become energized by the PV system. The current understanding of the codes and standards has led to the following problems:

- unsatisfactory module grounding measures
- violations of the module's UL 1703 listing because the installation does not comply with the installation manual's prescribed method of module frame grounding
- incorporation of components listed to more general grounding equipment standards that may or may not be suitable for the application, and/or
- well-engineered grounding means that have, at present, no clear path for demonstrating their adequacy to customers and inspectors (Ball 2011).

The two following issues were focused on in the report by Ball (2011).

The first is the lack of confidence in existing, approved grounding methods, which results from the many grounding failures observed in fielded systems. Although statistical studies of failure rates are not available, there is enough anecdotal evidence to support recommendations for additional testing and revision of standards.

The second major issue is the limited number of approved grounding methods and devices available for PV modules and systems that are certified or listed by nationally recognized testing laboratories. Industry stakeholders who would prefer to market or use new grounding methods and devices point out that the absence of certification for these products is not necessarily based on issues pertaining to safety or reliability but rather results from a lack of consensus in the assignment and development of applicable standards.

This interim report not only discusses the issues with grounding PV systems but also gives near term recommendations to improve the current codes and standards.

- Perform research testing to qualify the impact of different current levels in the continuity and component performance tests.
- Monitor and review developments during the revision of UL 467 to incorporate PV system-specific applications.
- Monitor and review results and developments from UL's enhanced environmental and corrosion resistance testing.
- Engage additional corrosion experts outside of the PV industry to help interpret the new test results and provide guidance on how they can be applied effectively in new or revised standards.
- Explore the possibility of developing special tests for coastal environments, again using guidance from other industries (such as the maritime industry) with relevant experience.

- Seek additional expertise on whether and how strain relief and force tests may be incorporated to evaluate grounding means based on the forces experienced during installation.
- Conduct additional research to identify and classify installation environments and to determine how they might impact grounding design, installation, and maintenance decisions (Ball 2011).

A report by Flicker and Johnson (2013) discusses ground faults and how blind spots associated with them can cause significant problems for PV systems.

A 2012 Solar America Board for Codes and Standards (Solar ABCs) publication, *The Ground-Fault Protection Blind Spot: Safety Concern for Larger PV Systems in the U.S.* (Brooks 2012), revealed that undetected faults on grounded PV array conductors were an initial step in a sequence leading to two well-publicized rooftop fires. In that paper, the theoretical detection limits of traditional ground fault protection systems were discussed but not explored in depth.

The specific wiring failure that occurred in both the Bakersfield and Mount Holly fires was that one of the conductors faulted but did not blow the protective ground-fault fuse. As a result, this established a new "normal condition" to be measured. Then when a second ground fault current occurred, the ground-fault protection device was unable to interrupt the current, allowing arc faults to be formed, spreading sparks to surrounding materials, causing ignition (Brooks 2012). The following quote from Brooks (2012) explains the similarities and differences between the two events:

Both reference fires show evidence of significant arcing at one location in the PV array, although the fire ignited in a completely different section of the array. That is, in both fires there was significant damage in a seemingly unrelated portion of the array away from the initial sources of ignition. Jackson connects the coincidence of these faults in his report, but he questions the likelihood of a repeat event. The *SolarPro* article showed that the problem was ultimately in the blind spot of ground-fault protection equipment and therefore an apparent general concern to the PV industry in the United States.

As in the Bakersfield Fire, the Mount Holly fire caused significant damage in two different locations at the same time. At Mount Holly, the faults existed inside two different combiner boxes at the same time. The likelihood of these two faults initiating simultaneously in different locations is extremely small. Thus, it is likely that one fault existed for some period of time prior to the initiation of the second one. This is the fundamental insight that led to an understanding of the root cause of the fires.

Understanding the root cause of these events should help prevent ground faults from starting a fire in the future. The Bakersfield fire occurred with a fault in a small ground conductor, which then became part of a circuit it was not designed for, while the Mount Holly Fire had both faults occur in the large combiner feed box. The National Electric Code (NEC) (NFPA 70) provides Ground-Fault Protection Device Test Standards that allow trip limits, the minimum amount of current that can be leaked before the Protection Device activates, i.e. "1 A for inverters rated up to 25 kW and 5 A for inverters greater than 250 kW". Any current leaked below these limits are within the "blind spot" of the system. The blind spot can cause issues because there could be a fault, such as with the Mount Holly or Bakersfield fires, but because it is not large enough to trip the Protection Device it can go unnoticed, and then cause a much more severe fault.

A report by Brooks (2012) focuses on the two fire incidents at Bakersfield and Mount Holly and provided the following preliminary mitigation strategies and equipment retrofit recommendations to reduce fire danger and prevent similar disasters in the future.

- proper installation techniques with close attention to wire management,
- annual preventative maintenance to identify and resolve progressive system damage,
- introduction to the use of data acquisition to monitor the operation of all PV systems at a level sufficient to determine if unscheduled maintenance is required, and
- additional ground-fault and PV array isolation sensing devices that can be incorporated into the data system to alert operators to potential problems so that maintenance personnel can be dispatched well in advance of damage that could lead to a fire.

The combination of proper installation, maintenance, monitoring and sensing devices are all strategies to be utilized for PV systems. Brooks (2012) discussed the important first steps of installing panels.

Conduct a detailed review of all installation-related issues and develop a punch list to address concerns, including wire management, grounding, and equipment installation for the entire system. Once the punch list has been resolved, the commissioning procedure should include:

- insulation resistance tests on all field-installed conductors, including modules and module wiring;
- open-circuit voltage and polarity tests on all string and feeder circuits;
- operational current readings on all series strings and feeders; and
- thermography of all inverters, disconnects, and combiner boxes at 50 % load or higher as well as thermography of the array to scan for hot spots not caused by shading or other normal temporary conditions.

Maintenance is the next element described by Brooks (2012):

Next, develop a maintenance schedule that establishes consistent inspection, documentation, and maintenance procedures to identify and correct problems before they result in a fire. The fact that a maintenance inspection prior to the Mount Holly fire identified conductor damage that was later recognized as a potential cause of the subsequent fire reveals the value of this type of inspection. This case also highlights the need to adequately train maintenance personnel so that they recognize the visual and testing indicators that a fire is possible. Maintenance procedures should include:

- visual inspection of all equipment and field connections in equipment for signs of damage or degradation;
- visual inspection of all accessible electrical junction boxes and raceways to see if conductors are damaged and in need of repair or replacement;
- visual inspection of string conductors to identify any physical damage that is in need of repair and additional protection to prevent progressive damage;
- operating voltage and current tests at defined conditions of irradiance and module temperature to compare output of strings;
- insulation resistance testing of modules, string wiring, and photovoltaic output circuits in the array (sometimes referred to as a "megger" test); and

- thermography of all inverters, disconnects, and combiner boxes at 50 % load or higher, as well as thermography of the array to scan for hot spots not caused by shading or other normal temporary conditions.

Protection devices are the last step to ensuring protection from ground faults and other associated ignition hazards. The Solar America Board for Codes and Standards is a leading a working group to research ignition hazards with PV systems. Although the research is not yet complete, they have already made some substantial conclusions, as described by Brooks (2012).

Early results from large PV systems retrofitted with protective devices indicate that these devices may eliminate the blind spot without requiring redesign of the system. The types of protective devices that have currently been retrofitted to existing inverters for evaluation include:

- differential current sensors (also known as residual current detectors or RCDs) installed on feeders entering the array combiners on each system, and
- insulation resistance monitors that measure the resistance to ground on a PV array while it is not operating. In addition, other protective equipment may help mitigate fire danger in new systems, including:
- contactor combiners, which constitute an additional safety step beyond the differential current sensors and insulation monitors included in new inverters;
- arc fault detectors, which are required by the 2011 National Electrical Code; and
- module-level controls, which can shut the power off from each module (Brooks 2012)

The report by Brooks (2012) discussed the theoretical detection limits of traditional ground fault protection systems, but they were not explored in depth. A new report by Flicker and Johnson (2013), "Analysis of Fuses for 'Blind Spot' Ground Fault Detection In Photovoltaic Power Systems" discusses this further providing reference to research done on the topic.

To further the analysis of ground fault protection in photovoltaic (PV) systems, scientists from Sandia National Laboratories developed a functional circuit model of the PV system including modules, wiring, switchgear, grounded or ungrounded components, and the inverter. This model was implemented using the Simulation Program with Integrated Circuit Emphasis (SPICE) modeling language. The Sandia Technical Report, *Photovoltaic Ground Fault and Blind Spot Electrical Stimulations* (Flicker and Johnson 2013), presents the complete derivation of the Sandia PV System SPICE model and the results of parametric fault current studies with varying array topologies, fuse sizes, and fault impedances. This Solar ABCs report contains the subsection of the Sandia technical report that focuses on blind spot ground faults to the grounded current-carrying conductor. The behavior of the array during these faults is studied for a range of ground fault fuse sizes to determine if reducing the size of the fuse improves ground fault detection sensitivity. Results of simulation studies show that reducing the amperage rating of the protective fuse does increase fault current detection sensitivity without increasing the likelihood of nuisance trips. However, this effect reaches a limit as fuses become smaller and their internal resistance increases to the point of becoming a major element in the fault current circuit (Flicker and Johnson 2013).

The following excerpt describes the best practices for creating a system that best detects ground faults and limits the blind spots associated with measurements.

While it may not be possible to provide complete detection for both faults within the array and faults to the grounded CCC using a fuse, the simulations indicate that the detection window for blind spot faults can be optimized by:

- minimizing leakage current, because fault current is the opposite direction of leakage current and large leakage currents will inhibit the detection of negative CCC faults;
- decreasing the fuse sizing for large arrays below UL 1741 requirements to 1 A, because module leakage current will be too small to result in nuisance tripping and it will trip on more ground faults;
- preventing the reduction in fuses below 1 A because the internal resistance of the fuse prevents the fault current from passing through the GFPD;
- monitoring both GFPD current magnitude and direction (especially for smaller array sizes), because GFPD current can change direction when a fault to the grounded CCC occurs; and
- employing other fault detection tools such as differential current measurement and insulation monitoring (see (Ball 2012, in press) for more information on alternative ground fault detection techniques and suggestions) (Flicker and Johnson 2013).[1]

PV system designers need to implement residual current monitors in their designs to augment the existing fuse-based detectors so that all ground faults can be detected. Arc fault detectors are now available in many string inverters on the market, but few arc detectors are available for use in larger central inverter systems. Designers must also specify these arc fault detectors in current designs so that PV systems meet the requirements of the 2011 and 2014 NEC 690.11.

---

[1]CCC refers to a "current carrying conductor".

Chapter 9
Electrical Hazards Associated with Fire Fighting Operations

9.1 Overview of Hazards

Due to the increase in popularity of PV systems, fire fighters, fire ground incident commanders, and other emergency first responders are encountering PV systems more often in fire events. "As a result of greater utilization, traditional fire fighter tactics for suppression, ventilation and overhaul have been complicated, leaving fire fighters vulnerable to potentially unrecognized exposure. Though the electrical and fire hazards associated with electrical generation and distribution systems is well known, PV systems present unique safety considerations" (Grant 2011). PV systems are a new hazard that is rapidly being introduced to fire fighting, but is also a complex hazard. Very few PV systems are the same, so consequently their fire situations may be unique.

Increasing magnitudes of exposure to electric current via an electrical shock will have the following effect on people:

- 0–2 mA, a person typically cannot even feel the shock
- 2.1–40 mA, person can perceive the shock and possibly be in pain.
- 40.1–240 mA, a person will lock on to the system and will lose muscle control.
- Greater than 240 mA, a person's heart could stop beating and possibly die.

It is difficult to generically assess how dangerous PV systems are and how much of a hazard they pose to fire fighters. By visually inspecting a PV system, the electrical charge cannot be determined within the panels, wires, conductors, and other elements of the PV system. Examples of the questions that are not easily answered following a simple visual inspection of a PV system are:

- Can contact with an exposed electrically charged element such as a PV panel electrocute a fire fighter?
- Can a wire or conduit be safely cut for ventilation?
- Can water by safely sprayed onto a fire that has electrically charged elements?

© Fire Protection Research Foundation 2015
R. Wills et al., *Best Practices for Commercial Roof-Mounted Photovoltaic System Installation*, SpringerBriefs in Fire, DOI 10.1007/978-1-4939-2883-5_9

- How can complete extinguishment of a fire be determined?
- How will a fire interact with the PV system? Will a PV system affect the fire growth rate or lead to an explosions?

Because these are difficult answers to obtain from a visual inspection of a PV system, a conservative approach has usually been taken by command officers. However, even though fire fighters have tried to take a conservative approach to their efforts, there are still many dangers from the unexpected. While the statistical data on fire fighter accidents are limited, Grant provides one statement to put this hazard into perspective:

> Statistical data indicates that on average 40,270 fire fighters were injured during fireground operations in the United States annually from 2003 through 2006. Of these injuries, there were on average 215 fire fighters engaged in fireground operation at a building fire whose injuries were due to "electric shock." Further, 50 of these annual injuries were considered moderate or severe injuries (Grant 2012).

To prevent these incidents from happening steps need to be taken preceding, during, and after an event to ensure the safety of fire fighters. A technical understanding of how PV systems affect fire fighters and their operations is fundamental to developing codes and training techniques. Quantifying those hazards and observing what happens in different situations can lead to improved measures for protecting emergency responders from the hazards. The improved understanding can be done through studies and experiments, the start of which has been occurring in recent years.

9.2 Mitigation Measures

Guidelines, codes, and fire fighter training activities have been developed recently to address electrical hazards with PV systems. In order to increase public safety for all structures equipped with PV systems, the California Department of Forestry and Fire Protection produced a guideline in 2008 called "Solar Photovoltaic Installation Guideline" California (2008). The guideline gives instruction for the design, assembly, and installation of PV systems so that the objectives of the solar PV industry and the fire service can be achieved. The recommended practices include marking, access, pathways, smoke ventilation, and location of DC conductors. These guidelines were developed by fire service officials, as well as solar industry experts. The collaboration between the two parties created a guideline that incorporated diverse set of experiences and backgrounds.

The guidelines created by California (2008) were eventually revised and adopted into Section 605 of the 2012 edition of the International Fire Code (IFC). Brooks (2011) emphasizes the importance of understanding the guideline given that the guideline has transformed from a best practice standard to a legally binding code. Brooks (2011) explains that the guideline that was originally produced provides specific information as to what the exact requirements are for the PV systems.

Brooks explains the logic behind the guideline. Without this explanation local fire officials may not be able to fully understand how the code can be applied and more importantly how it can be altered to fit a certain situation (Brooks 2011). For example, pathways between PV systems are required for commercial buildings. They are required from the outside perimeter access areas to ventilation areas such as skylights. Brooks (2011) states, "These pathways ensure that the fire fighter is free to move around the perimeter and to access the ventilation location closest to the fire." As such, if two ventilation areas are relatively close together, then these nearby ventilation areas may be considered one entity such that the requirements may not have to be duplicated in each ventilation area.

The above requirements for PV systems on roofs are addressed in Section 11.12 Photovoltaic Systems of NFPA 1. These provisions include marking requirements for the main service disconnect, circuit disconnecting means, conduit and cable assembly, secondary power source, and inverters. Provisions for access, pathways, and smoke ventilation are also included. Minimum distances of the array from the edges of the roof are described so that a fire fighter will have room to properly access the fire safely. The locations of the direct current (dc) conductors are also detailed to provide maximum ventilation opportunities and minimize trip hazards. These are the most currently adopted requirements that should be used in the installation of PV systems.

Two documents by Grant and CAL FIRE were developed in 2010 that described proper fire fighting tactics with PV systems. These two reports focused more on best practices during an emergency incident as opposed to the previously mentioned documents that focused on mitigation strategies before an incident occurred. The main goal of a report by Grant (2010) was to assemble and widely disseminate core principle and best practice information for fire fighters, fire ground incident commanders, and other emergency first responders to assist in their decision making process at emergencies involving solar power systems on buildings. CAL FIRE (2010) also published a training document for fire operations during photovoltaic emergencies. The best practices for fire fighters mentioned in these reports are noted in the following sections.

Important mitigation measures that have been taken are the changes to the NEC. The 2014 edition of the NEC® has incorporated changes to address fire fighter safety concerns that arose from the Target fire in Bakersfield, as well as some other incidents. One of the key features is new Section 690.12 on rapid shutdown, which states:

690.12 Rapid Shutdown of PV Systems on Buildings.
PV system circuits installed on or in buildings shall include a rapid shutdown function that controls specific conductors in accordance with 690.12(1) through (5) as follows.

(1) Requirements for controlled conductors shall apply only to PV system conductors of more than 1.5 m (5 ft) in length inside a building, or more than 3 m (10 ft) from a PV array.

(2) Controlled conductors shall be limited to not more than 30 volts and 240 volt-amperes within 10 seconds of rapid shutdown initiation.

(3) Voltage and power shall be measured between any two conductors and between any conductor and ground.

(4) The rapid shutdown initiation methods shall be labeled in accordance with 690.56(B).
(5) Equipment that performs the rapid shutdown shall be listed and identified (NEC 2014).

The NEC does not specify which equipment must perform the rapid shutdown function. The shutdown could be at the combiner box, at a module level dc-dc converter (which may be a power optimizer), at single module micro-inverter, or at the module itself. However, rapid shutdown requirements are contrary to a key characteristic of PV systems, i.e. PV systems are almost always active.

Another provision in the NEC is for disconnecting means for utility-interactive inverters that are mounted in not readily accessible locations.

Utility-interactive inverters shall be permitted to be mounted on roofs or other exterior areas that are not readily accessible and shall comply with 690.15(A)(1) through (4):

(1) A dc PV disconnecting means shall be mounted within sight of or in each inverter.
(2) An ac disconnecting means shall be mounted within sight of or in each inverter.
(3) The ac output conductors from the inverter and an additional ac disconnecting means for the inverter shall comply with 690.13 (A).
(4) A plaque shall be installed in accordance with 705.10 (NEC 2014).

In the 2014 edition of the NEC, the different types of disconnects are defined so that specific requirements for each type can be described;

690.17 Disconnect Type.
(D) Manually Operable
The disconnecting means for ungrounded PV conductors shall consist of a manually operable switch(es) or circuit breaker(s). The disconnecting means shall be permitted to be power operable with provisions for manual operation in the event of a power-supply failure. The disconnecting means shall be one of the following listed devices:

(1) A PV industrial control switch marked for use in PV systems
(2) A PV molded-case circuit breaker marked for use in PV systems
(3) A PV molded-case switch marked for use in PV systems
(4) A PV enclosed switch marked for use in PV systems
(5) A PV open-type switch marked for use in PV systems
(6) A dc-rated molded-case circuit breaker suitable for backfeed operation
(7) A dc-rated molded-case switch suitable for backfeed operation
(8) A dc-rated enclosed switch
(9) A dc-rated open-type switch
(10) A dc-rated rated low-voltage power circuit breaker

(B) Simultaneous Opening of Poles
The PV disconnecting means shall simultaneously disconnect all ungrounded supply conductors.
(C) Externally Operable and Indicating
The PV disconnecting means shall be externally operable without exposing the operator to contact with live parts and shall indicate whether in the open or closed position.
(D) Disconnection of Grounded Conductor
A switch, circuit breaker, or other device shall not be installed in a grounded conductor if operation of that switch, circuit breaker, or other device leaves the marked, grounded conductor in an ungrounded and energized state (NEC 2014).

The required wiring methods are also described in the NEC so that wiring can be identified and grouped properly.

(B) Identification and Grouping

PV source circuits and PV output circuits shall not be contained in the same race-way, cable tray, cable, outlet box, junction box, or similar fitting as conductors, feeders, branch circuits of other non-PV systems, or inverter output circuits, unless the conductors of the different systems are separated by a partition. PV system conductors shall be identified and grouped as required by 690.31(B)(1) through (4). The means of identification shall be permitted by separate color coding, marking tape, tagging, or other approved means.

(1) PV Source Circuits

PV source circuits shall be identified at all points of termination, connection, and splices.

(2) PV Output and Inverter Circuits

The conductors of PV output circuits and inverter input and output circuits shall be identified at all points of termination, connection, and splices.

(3) Conductors of Multiple Systems

Where the conductors of more than one PV system occupy the same junction box, raceway, or equipment, the conductors of each system shall be identified at all termination, connection, and splice points.

Exception: Where the identification of the conductors is evident by spacing or arrangement, further identification shall not be required.

(4) Grouping

Where the conductors of more than one PV system occupy the same junction box or raceway with a removable cover(s), the ac and dc conductors of each system shall be grouped separately by cable ties or similar means at least once and shall then be grouped at intervals not to exceed 1.8 m (6 ft).

Exception: The requirement for grouping shall not apply if the circuit enters from a cable or raceway unique to the circuit that makes the grouping obvious (NEC 2014).

9.2.1 Electrically Active PV Systems

A best practice is to assume that all PV systems are active. PV Systems on large commercial buildings produce significant amounts of electricity and fire fighters need to take caution. Standard Operating Procedures (SOPs) are not available for PV systems specifically; it is normally grouped into electrically charged components. Size-up strategies and tactics are determined based on both training and a well thought out approach to the problem. The importance of recognizing the PV system and how it is integrated with the building is important to understand before attacking the fire. Knowing that a PV system is likely to be electrically charged should affect the way fire fighters will fight the fire. The approach taken is conservative, with the intent of limiting the spread of the fire and not engaging the PV system.

Captain Matt Paiss of the San Jose Fire Department has released multiple YouTube videos that describe these proper fire fighting strategies associated with PV systems. Paiss (2011a) describes how to determine if there is a PV system involved with the fire by looking for labeling such as that illustrated in Fig. 9.1.

Fig. 9.1 Labeling for PV
System (Paiss 2011a)

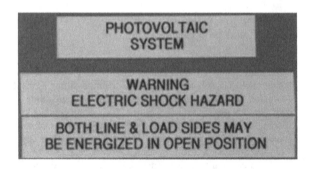

The label illustrated in Fig. 9.1 is actually for a residential PV system. New requirements will require increased labeling of systems so that fire fighters know what if PV systems are present. The two YouTube videos by Paiss include additional strategies such as isolation from the system and delivering power back into the grid. Showing examples of what these systems look like and how they operate through conduit and inverters help fire fighters understand what hazards are posed to them and how they can protect themselves from those hazards.

The following excerpt from Grant (2013) summarizes the basic precautions and actions that should be taken by fire fighters if there is an active PV system.

Certain basic safety precautions should be taken into account by all fire fighters on the fireground. Determining the presence of a PV system is key to preventing fireground injuries.

The following six points of safe operation are offered for fire fighters:

- Daytime = Danger; Nighttime = Less Hazard
- Inform the IC that a PV system is present
- Securing the main electrical does not shut down the PV modules
- At night apparatus-mounted scene lighting may produce enough light to generate an electrical hazard in the PV system
- Cover all PV modules with 100% light-blocking materials to stop electrical generation
- Do not break, remove, or walk on PV modules, and stay away from modules, components, and conduit

A photovoltaic array will always generate electricity when the sun shines. These units do not turn "off" like conventional electrical equipment. Fire fighters on the fireground should always treat all wiring and components as energized. Breaking or compromising a photovoltaic module is extremely dangerous and could immediately release all the electrical energy in the system.

Without light, photovoltaic panels do not generate electricity, and thus nighttime operations provide less of a hazard. Emergency scene lighting during a nighttime fireground operation, such as from a mobile lighting plant unit, or sources other than direct sunlight, may be bright enough for the photovoltaic system to generate a dangerous level of electricity.

In summary, there are several fundamental points of consideration for fire fighters and incident commanders when handling any building fire equipped with a solar power system:

- Identify the existence of a solar power system

    - locate rooftop panels
    - clarify electrical disconnects
    - obtain system information

- Identify the type of solar power system

  - Solar Thermal System
  - Photovoltaic System

- Isolate and shutdown as much of the system as possible

  - Lock-out and tag-out all electrical disconnects
  - Isolate the photovoltaic system at the inverter using reliable method

- Work around all solar power system components

While salvage covers can be used to block sunlight, some electricity will still be generated unless they are made of material that is 100% light blocking. Care is needed to make sure that wind does not suddenly blow off any salvage covers covering panels. Foam is not effective in blocking sunlight, and will slide off the solar array.

Fire fighters are routinely exposed to potentially life-threatening risks. When these risks are properly understood, the value of putting a fire fighter at risk can be properly managed based upon the benefit at the fire scene. The recent media coverage from the Dietz and Watson fire emphasized this issue as shown in the following quotes:

More than 7,000 solar panels on the roof of a burning warehouse in Burlington County proved too much of a hazard for fire fighters, local officials said today.

'We may very well not be able to save buildings that have alternative energy,' William Kramer, New Jersey's acting fire marshall [sic], said after Delanco Fire Chief Ron Holt refused to send his fire fighters onto the roof of the 300,000-square foot Dietz & Watson facility, ablaze since Sunday afternoon.

Solar panels are particularly hazardous to fire fighters for a number of reasons, according to Ken Willette, a division manager with the National Fire Protection Association.

'There is a possibility of electric shock because the electricity to the panels can't be shut off,' he said, 'and not having a clear path on the roof to cut a ventilation hole is another challenge' (Augenstein).

While these quotes underline concerns that the fire service has when fighting a fire, a balanced view of this fire event would put all the hazards, including those related to the PV system, in the proper context of the fire scene. This is why media reports are of limited benefit in the literature review. Each reporter has a perspective they are trying to communicate to their readership, whereas a detailed fire investigation report should be much more balanced in the treatment of the hazard issues.

In 2011, UL (Backstrom 2011) did an in depth study of multiple factors that led to the harm of fire fighters in PV system fires. The single factor that affected all other points of their research was that PV systems cannot be powered down. They can be turned off in the disconnect box, but the panels themselves remain electrically charged as long as there is light for them to convert into energy. Contrary to the assumption by some fire fighters that the panels are safe to handle at night because it is dark, UL's research found that even in the dark the lighting used by the fire service can cause PV panels to fully charge (Backstrom 2011).

The research also found that using a tarp to cover the panels fully is not a good practice to de-energize the PV system. Class A foam in both 0.5 and 1.0 %

concentrations that were sprayed onto the panels in testing were also proven to not be very effective. Not only are partially blocked panels still active, but also damaged panels can still be active. Even when damaged in a full house fire, UL's research found that the majority of the PV panels were still fully functional even though damaged. This means that the panels were still able to fully charge, causing electrical hazards, and also provided the extra danger of loose wires and current through other metal elements.

To determine the efficiency of clothing protection for fire fighters from being shocked, UL also tested three types of gloves and boots, including both new and old equipment having various degrees of wear and tear. As a result of these tests, UL found that new firefighting gloves and boots tended to provide sufficient electrical insulation. Yet as they went through wear and tear, soiled boots and gloves changed to being good conductors of electricity, especially as gloves became wet or the metal heels or toes of the boots were exposed to charged materials. As such, this equipment is not considered equivalent to electrical personal protective equipment.

9.2.2 Suppression Tactics

Since a current can travel through water/foam to a fire fighter using the suppression device, sprays of suppression agents may become electrically charged components. Unlike other electrical equipment, because PV systems cannot be de-energized, the traditional suppression techniques need to be altered. Traditional fire suppression training discourages the use of water on energized circuits. However, at a safe distance and with the correct hose stream, water may be the most effective fire suppression method.

UL's research on how to best protect fire fighters from electrical hazards included testing water nozzle sprays; gloves and boots. These tests used different voltage, diameter of nozzle, spray pattern, distance from source to nozzle, water conductivity, and water pressure to determine leakage current in mA. The different types of nozzles used were smooth bore aluminum, "with three stacked tips of 1, 1-1/8 and 1-1/4 in.", and adjustable which could be adjusted "from a solid stream to a wide fog". The main finding from these tests was that it was very difficult to sustain dangerous leakage currents. At 1000 Vdc only a solid water stream at a distance of up to 10 ft could carry current (Backstrom 2011). Longer distances or going to a narrow fog pattern eliminated the current flow. It was also discovered that hose streams with a higher pressure tend to break up more and thus produce less leakage current.

Research is needed to assess whether outdoor weather exposure rated electrical enclosures are resistant to water penetration by fire hose streams. A typical enclosure will collect water and present an electrical hazard (Backstrom 2011).

An additional area of research is to develop guidance on when complete extinguishment of a PV panel fire is achieved. A particular concern is the assessment of the potential extinguishment of a PV panel during a nighttime fire. Once the

PV panel is exposed to sunlight, it could begin to generate electricity again and create a shock hazard or re-kindle the fire Grant (2010).

9.2.3 Ventilation Tactics

Fire fighters may be on top of roofs in order to carry out ventilation tactics and remove smoke or heat from inside the structure. The introduction of PV systems onto roofs limits their ability to maneuver on the roof as well as they are exposed to electrical hazards of being shocked by an active panel or wire. The design of the PV system directly affects how it behaves during an emergency situation and how the fire fighters can manage the PV system. The following guidelines and codes address how to design a system so that a fire fighter can perform their duties of venting and suppressing the fire.

The California (2008) guideline addresses the need for appropriate marking of all PV systems. However, fire fighters cannot be assumed to have an understanding of PV systems and which wires and components carry a high current. "This can facilitate identifying energized electrical lines that connect the solar modules to the inverter, as these should not be cut when venting for smoke" (California 2008). The marking allows them to be able to determine what areas are safe for their procedures and which areas are not.

PV systems are addressed in Section 605.11 of the IFC, which are a revised version of California's (2008) guidelines. The IFC requires that the systems shall be installed in accordance with Sections 605.1.1 through 604.11.4, the IBC, and NFPA 70. Section 605.11.1 discusses marking requirements on interior and exterior direct-current (DC) conduit, enclosures, raceways, cable assemblies, junction boxes, combiner boxes and disconnects. There is a uniform marking style and marking content. The materials used for marking must be reflective and weather resistant. The marking is required to be placed adjacent to the main service disconnect in a clearly visible area. The marking is required every 10 ft (3048 mm), within 1 ft (305 mm) of turns or bends and within 1 ft (305 mm) above and below penetrations of roof/ceiling assemblies, walls or barriers.

The next section of the IFC, Section 605.11.2, requires conduit, wiring systems, and raceways for photovoltaic circuits to be located as close as possible to the ridge or hip or valley and from the hip or valley as directly as possible to an outside wall to reduce trip hazards and maximize ventilation opportunities. Conduit runs between sub-arrays and to DC combiner boxes must be installed in a manner that minimizes the total amount of conduit on the roof by taking the shortest path from the array to the DC combiner box. DC combiner boxes must be located such that conduit runs are minimized in the pathways between arrays. DC wiring must be installed in metallic conduit or raceways when located within enclosed spaces in a building. Conduit must run along the bottom of load bearing members.

Sections 605.11.3.1 through 605.11.3.3.3 of the IFC discuss roof access, pathways, and spacing requirements for residential and commercial buildings. Due to the larger size of commercial buildings, the requirements are subsequently more extensive. For example, for a residential building the PV system cannot be higher than 3 ft from the ridge of a house, and must have a three-foot wide access pathway. For commercial buildings, an access pathway must be a minimum of 8 ft in width for smoke ventilation purposes. These pathways for commercial buildings are required to be capable of supporting the live load of fire fighters accessing the roof. These pathways may also not be located where a ground ladder would have to be placed over openings such as windows, and must not conflict with overhead obstructions such as tree limbs, wires or signs.

Additional requirements in Section 605.11.3 of the IFC address the arrangement of the pathways. Centerline axis pathways are to be provided in both axes of the roof, and a four-foot (1290 mm) clear pathway must be provided for ventilation hatches. For smoke ventilation purposes there are additional requirements for the pathways depending on the arrangement of the PV system. For example, PV arrays can be no larger than 150 ft (45,720 mm) by 150 ft (45,720 mm) in distance in either axis. Between these PV arrays a pathway of 8 ft (2,438 mm) wide is required, with a 4 ft (1,290 mm) wide pathway provided to skylights or smoke and heat vents. A 4 ft (1,290 mm) wide pathway is allowed when ventilation cut outs of 8 ft by 4 ft (2,580 mm by 1,290 mm) are provided every 20 ft (6,096 mm) on alternating sides of the pathway.

The 2012 IFC requirements take a proactive approach to dealing with the electrical hazards posed to fire fighters. By having labels on the equipment, it will allow personnel to make an educated assessment of the hazard. The pathways allow for areas for fire fighters to walk on the roofs as well as provide vertical ventilation to the fire. Requirements for wiring reduce the trip/fall hazard for the fire fighters as well.

Chapter 10
Weather-Related Maintenance Considerations

The manufacturer warranty period typically exceeds 20 years for crystalline silicon modules and 15 years for thin-film modules. Unfortunately, there is little or no systematically field monitored data or independent accelerated test data available to support most of these warranty claims. Investors, financiers, power purchasing agreement companies, and consumers are now expecting objective substantiations for these warranty claims. The PV module components, including cells and polymeric materials, must be protected from degradative losses (soft/durability losses) and catastrophic failures (hard/reliability failures) caused by stresses including temperature, humidity, ultraviolet (UV) radiation, wind, hail, and high system voltages, as well as effects including corrosion, broken interconnects, hotspots, delamination, and encapsulant discoloration (TamizhMani and Kuitche 2013).

The rating of the roof is affected during both the installation of the PV systems as well as during throughout the lifetime of the PV system. If the PV system was not installed correctly then it could lead to more severe maintenance problems. Tampering with the roof structure can cause wear and tear over time and could further decrease the rating of the roof. In response to these issues, the Center PV Taskforce created a document that identifies four main objectives when installing PV panels in order to minimize their negative impact on the roof.

The following four objectives cover different hazards that are introduced to the roof when installing a PV system.

- PV module attachments and workers should not compromise the roof's water proofing integrity, especially through transferring loads in a way that avoids overloading or damaging the roof system.
- Every part connected to or used by the PV module should be designed for performance, durability and service life equivalent to the expected PV module service life. The entire system could be compromised if any element does not meet the criteria used for the entire PV system.

© Fire Protection Research Foundation 2015
R. Wills et al., *Best Practices for Commercial Roof-Mounted Photovoltaic System Installation*, SpringerBriefs in Fire, DOI 10.1007/978-1-4939-2883-5_10

- Water needs to be able to drain correctly from the roof during and after PV installation. Siting pools of water can degrade roofing and the systems components as well.
- Provide accessibility to the roof and PV system design to allow for effective inspection, maintenance and repair.

By following these guidelines, PV systems and the roofs they are installed on can keep their fire resistance integrity, therefore protecting the building they are on and maximizing their production potential.

Chapter 11
Compilation of Best Practices

11.1 Structural Loading

Guidelines depend on what type of mounting is used to attach the PV systems to the roof. There are three different methods of mounting PV systems to a roof structure: ballast-only, attached roof-bearing, and structurally attached. The attachment method can significantly affect the loads that are being applied to the structure and how it is being handled. "The roofing industry has learned from experience that ballast-only rooftop equipment does not necessarily remain stationary. Structurally attached equipment is more reliable in this regard" (Kirby 2011). An engineer using calculations found in codes and standards can evaluate structurally attached equipment. For example, wind loads can be determined using *ASCE Standard 7-05*, the standard for evaluating wind forces on structures. In addition to wind loads, other loads such as snow, seismic and gravity (dead load) must be taken into account" (O'Brien and Banks 2012). The load that is provided by the weight of the PV systems themselves is only a portion of the loads that is going to be imposed on the roofing structure.

11.2 Wind Loads

There are two methods to determine the wind loads that a PV system will be exposed to: wind tunnel testing and calculation. Although wind tunnel testing can demonstrate the effect of wind loads on PV systems, such testing may not be the most feasible option for every individual PV system.

> Wind tunnel testing can provide an appropriate basis for design of rooftop solar arrays per the code if the testing is done properly and the results of these tests are properly applied (Kopp et al. 2013). However, conducting this method can be very complex and expensive. Few facilities can meet the minimum requirements of this method. Because the cost, time

© Fire Protection Research Foundation 2015
R. Wills et al., *Best Practices for Commercial Roof-Mounted Photovoltaic System Installation*, SpringerBriefs in Fire, DOI 10.1007/978-1-4939-2883-5_11

and effort required to perform this type of testing for each specific PV project would be prohibitive, the challenge is to develop a set of test data that can be used to provide design loads for a wide variety of different buildings, sites and array shapes. This type of generalization is possible with the appropriate test program, but is a complex and challenging undertaking (Kopp et al. 2013).

Using dynamic calculations is a better way to determine the wind loads a roof mounted PV system would be exposed to. An advanced system for calculating wind loads is described in the SEAOC Report PV2-2012. In this document, an equation is edited to incorporate more factors such as the solar panel height above the roof and the low edge and raised edge, chord length of the solar panel, width of the overall building, parapet height factor, array edge increase factors, among others.

The SEAOC method is the most accurate and cost effective way available to calculate wind forces on PV systems. There is still progress to be made with this system to make it more specific for each situation. A future goal is to have the ideas and methods presented in the report be adopted by the International Code Council. Wind design of roof-mounted PV systems will be addressed in ASCE 7 in 2016 and then possibly by the IBC in 2018. Although this resource is the most effective way to calculate wind forces, it still has its limitations. Structurally attached and ballast-only PV systems are the only attachment methods that have a determined way of predicting wind forces in every direction. Many PV systems are not structurally attached and need an alternate method to predict these wind forces. Also, deflectors/shrouds are not addressed within SEAOC Report PV2-2012.

11.3 Hail

The best way to prevent a PV system failure from hail is to have the system tested and approved through a standard testing procedure. Hail hazards are addressed by a set of FM Global's Approval Standards. There are two types of modules that FM Global refers to: rigid modules and flexible modules. As well as FM, there are other standards from the following two organizations are most commonly used for PV systems.

ASTM E1038 is a testing standard that is used to determine the resistance of PV systems to hail. This standard uses propelled ice balls to simulate hailstones. The effects of impact may be either physical or electrical degradation of the module. The testing standard specifies the proper method for mounting the test specimen, conducting the impact test, and reporting the effects. The mounting method tested depends on the arrangement that will be used in actual installations. Different impact locations are determined based on vulnerable areas on the array. The size and weight of the ice balls are also specified. The velocities of the ice balls are meant to be comparable to speeds that hailstones could hit a PV system. The ASTM E1038 standard does not establish pass or fail levels but instead provides a procedure for determining the ability of photovoltaic modules to withstand impact forces of falling hail.

In Section 30 of UL 1703 "Flat-Plate Photovoltaic Modules and Panels," the "Impact Test" is described that is used on panels submitted for listing. Within this testing guideline it states that there may be no particles larger than 1 in.^{2} (6.5 cm^2) released from their normal mounting position. A 2 in. steel ball is used to represent a hailstone falling onto the panel. As in ASTM E1038, the mounting of the PV system is to be representative of its intended use. Other testing procedures are described such as the distance that the ball must fall from, the location of the impact, among others that ensure that the test is representative of an actually hailstone striking the panel. The ASTM E1038 standard does not establish pass or fail levels but instead provides a procedure for determining the ability of photovoltaic modules to withstand impact forces of falling hail.

11.4 Snow

An assessment of possible snowdrift build-up should be considered in the construction and installment of PV systems in areas where snow is expected (The Center PV Taskforce 2012). In order to provide a complete assessment, snow loads could be calculated in a similar manner as wind loads. However, documentation of this type of analysis is limited. Snow loads are addressed in the IEC 61215. An accelerated stress test using static mechanical loading for PV systems represents loads that snow buildup would introduce. On a related issue, the IEC 61215 also has a salt spray test that is used for determining corrosion due to salt used for removal of snow and ice.

11.5 Debris Accumulation

"PV Racking and Attachment Criteria for Effective Low-Slope Roof System Integration" by The Center PV Taskforce has recommendations for five fundamental principles of effective roof system integration; external forces, system integration, roof drainage, roof and PV system maintenance, and roof safety. The roof drainage section describes solutions for preventing and dealing with sitting roof water and debris accumulation. For example, limiting horizontal elements, providing accessible roof drains for periodic maintenance, and providing walkway areas for roof inspection and maintenance all help prevent debris build up.

11.6 Seismic

SEAOC has a draft document that addresses the seismic hazards associated with rooftop PV systems: "Structural Seismic Requirements and Commentary for Rooftop Solar Photovoltaic Arrays" (SEAOC Report PV1-2012).

For each of the three attachment methods there are separate requirements:
Fully framed systems:

PV support systems that are attached to the roof structure shall be designed to resist the
lateral seismic force F_p specified in ASCE 7-05 Chapter 13 (SEAOC Report PV1-2012).

Attached roof-bearing systems:

For attached roof-bearing systems, friction not to exceed $(0.9 \, \mu_s - 0.2 S_{DS}) W_{pf}$, is permitted
to resist the lateral force F_p where W_{pf} is the component weight providing normal force at
the roof bearing locations, and μ_s is the coefficient of friction at the bearing interface. The
resistance from friction is permitted to contribute in combination with the design lateral
strength of attachments to resist F_p (SEAOC Report PV1-2012).

Unattached (ballast-only):

Unattached (ballast-only) systems are permitted when all of the following conditions are
met:

- The maximum roof slope at the location of the array is less than or equal to 7 degrees
  (12.3 percent).
- The height above the roof surface to the center of mass of the solar array is less than the
  smaller of 36 inches and half the least plan dimension of the supporting base of the
  array.
- The system is designed to accommodate the seismic displacement determined by one of
  the following procedures:

  - Prescriptive design seismic displacement
  - Nonlinear response history analysis
  - Shake table testing (SEAOC Report PV1-2012).

These requirements and minimum separation requirements to allow certain
design seismic displacements are described further in SEAOC Report PV1-2012.

Along with prescriptive requirements, testing can be used to determine the
strength of the PV system against seismic hazards. Testing procedures are described
further in the SEAOC Report PV1-2012.

11.7 Fire Hazards

11.7.1 Flammability of Components

The flammability characteristics of the components that make up a PV system
depend on two factors: the materials used, and the arrangement of those materials.
As described before, mounting devices that are used to achieve a certain angle
towards the sun create gaps between the PV panels and the roof. Different materials
are used to support the PV modules themselves depending on the manufacturer and
installer. PV panels themselves do not represent the flammability characteristics of
an entire PV system that is attached to a roof. Rated PV systems are now required,
which include the panels, the attachment method, and the roofing structure.

It is difficult to apply this requirement of an entirely rated system incorporating both the roof and PV system. PV systems can be so unique that there is no simple way to determine the fire resistance rating. The testing that has been done previously only covers a limited number of PV systems. Before 2013, there wasn't a standard that covered how to conduct fire resistance tests. UL 1703 was recently updated to include testing details that reflect those used in prior research efforts (Sherwood et al. 2013). Having testing procedures available is a significant improvement for the PV industry. This will give designers a better ability to determine the fire rating of a PV system and how it affects the fire resistance rating of a roof. One concern about this approach is that most PV systems are unique and testing every single situation could become expensive. Testing needs to be reasonable or else a drop in PV system installation could be expected. Therefore the current best practice is to follow the testing standards within UL 1703 with the most exact representation of the PV system that is feasible.

11.7.2 Ignition Hazards

The PV panels themselves can be combustible as well as the components that make up the entire system. Electrically charged components such as the panels and wiring also present an additional hazard when introduced to fire. Early component failure, especially in wiring, leading to an electrical fault, is often the source and the leading cause of PV system building fires. There are many best practices that are described in reports by Brooks. A report by Brooks (2012) focuses on the two fire incidents at Bakersfield and Mount Holly and provided the following preliminary mitigation strategies and equipment retrofit recommendations to reduce fire danger and prevent similar disasters in the future.

- proper installation techniques with close attention to wire management,
- annual preventative maintenance to identify and resolve progressive system damage,
- introduction to the use of data acquisition to monitor the operation of all PV systems at a level sufficient to determine if unscheduled maintenance is required, and
- additional ground-fault and PV array isolation sensing devices that can be incorporated into the data system to alert operators to potential problems so that maintenance personnel can be dispatched well in advance of damage that could lead to a fire.

The combination of proper installation, maintenance, monitoring and sensing devices are all strategies to be utilized for PV systems. Brooks (2012) discussed the important first steps of installing panels.

Conduct a detailed review of all installation-related issues and develop a punch list to address concerns, including wire management, grounding, and equipment installation for the entire system. Once the punch list has been resolved, the commissioning procedure should include:

- insulation resistance tests on all field-installed conductors, including modules and module wiring;
- open-circuit voltage and polarity tests on all string and feeder circuits;

- operational current readings on all series strings and feeders; and
- thermography of all inverters, disconnects, and combiner boxes at 50 % load or higher as well as thermography of the array to scan for hot spots not caused by shading or other normal temporary conditions.

Maintenance is the next element described by Brooks (2012):

Next, develop a maintenance schedule that establishes consistent inspection, documentation, and maintenance procedures to identify and correct problems before they result in a fire. The fact that a maintenance inspection prior to the Mount Holly fire identified conductor damage that was later recognized as a potential cause of the subsequent fire reveals the value of this type of inspection. This case also highlights the need to adequately train maintenance personnel so that they recognize the visual and testing indicators that a fire is possible. Maintenance procedures should include:

- visual inspection of all equipment and field connections in equipment for signs of damage or degradation;
- visual inspection of all accessible electrical junction boxes and raceways to see if conductors are damaged and in need of repair or replacement;
- visual inspection of string conductors to identify any physical damage that is in need of repair and additional protection to prevent progressive damage;
- operating voltage and current tests at defined conditions of irradiance and module temperature to compare output of strings;
- insulation resistance testing of modules, string wiring, and photovoltaic output circuits in the array (sometimes referred to as a "megger" test); and
- thermography of all inverters, disconnects, and combiner boxes at 50 % load or higher, as well as thermography of the array to scan for hot spots not caused by shading or other normal temporary conditions.

Protection devices are the last step to ensuring protection from ground faults and other associated ignition hazards. The Solar America Board for Codes and Standards is a leading a working group to research ignition hazards with PV systems. Although the research is not yet complete, they have already made some substantial conclusions, as described by Brooks (2012).

Early results from large PV systems retrofitted with protective devices indicate that these devices may eliminate the blind spot without requiring redesign of the system. The types of protective devices that have currently been retrofitted to existing inverters for evaluation include:

- differential current sensors (also known as residual current detectors or RCDs) installed on feeders entering the array combiners on each system, and
- insulation resistance monitors that measure the resistance to ground on a PV array while it is not operating. In addition, other protective equipment may help mitigate fire danger in new systems, including:
- contactor combiners, which constitute an additional safety step beyond the differential current sensors and insulation monitors included in new inverters;
- arc fault detectors, which are required by the 2011 National Electrical Code; and
- module-level controls, which can shut the power off from each module Brooks (2012).

The report by Brooks (2012) discussed the theoretical detection limits of traditional ground fault protection systems, but they were not explored in depth. A new report by Flicker and Johnson (2013), "Analysis of Fuses for 'Blind Spot' Ground Fault Detection In Photovoltaic Power Systems" discusses this further providing reference to research done on the topic. The following excerpt describes the best practices for creating a system that best detects ground faults and limits the blind spots associated with measurements.

While it may not be possible to provide complete detection for both faults within the array and faults to the grounded CCC using a fuse, the simulations indicate that the detection window for blind spot faults can be optimized by:

- minimizing leakage current, because fault current is the opposite direction of leakage current and large leakage currents will inhibit the detection of negative CCC faults;
- decreasing the fuse sizing for large arrays below UL 1741 requirements to 1 A, because module leakage current will be too small to result in nuisance tripping and it will trip on more ground faults;
- preventing the reduction in fuses below 1 A because the internal resistance of the fuse prevents the fault current from passing through the GFPD;
- monitoring both GFPD current magnitude and direction (especially for smaller array sizes), because GFPD current can change direction when a fault to the grounded CCC occurs; and
- employing other fault detection tools such as differential current measurement and insulation monitoring (see [Ball 2012, in press] for more information on alternative ground fault detection techniques and suggestions) (Flicker and Johnson 2013).

PV system designers need to implement residual current monitors in their designs to augment the existing fuse-based detectors so that all ground faults can be detected. Arc fault detectors are now available in many string inverters on the market, but few arc detectors are available for use in larger central inverter systems. Designers must also specify these arc fault detectors in current designs so that PV systems meet the requirements of the 2011 and 2014 NEC 690.11.

11.8 Electrical Hazards Associated with Fire Fighting Operations

Due to the increase in popularity of PV systems, fire fighters, fire ground incident commanders, and other emergency first responders are encountering them more often in fire events. To prevent incidents from happening steps need to be taken preceding, during, and after an event to ensure the safety of fire fighters.

Guidelines, codes, and fire fighter training activities have been developed recently to address electrical hazards with PV systems. In order to increase public safety for all structures equipped with PV systems, the California Department of Forestry and Fire Protection produced a guideline in 2008 called "Solar Photovoltaic Installation Guideline" California (2008). The guideline gives instruction for the design, assembly, and installation of PV systems so that the objectives of the solar PV industry and the fire service can be achieved. The recommended practices include marking, access, pathways, smoke ventilation, and location of DC conductors. These guidelines were developed by fire service officials, as well as solar industry experts. The collaboration between the two parties created a guideline that incorporated diverse set of experiences and backgrounds.

The guidelines created by California (2008) were eventually revised and adopted into Section 605 of the 2012 edition of the IFC. Brooks (2011) emphasizes the importance of understanding the guideline given that the guideline has transformed from a best practice standard to a legally binding code. Brooks (2011) explains that the guideline that was originally produced provides specific information as to what the exact requirements are for the PV systems.

Brooks explains the logic behind the guideline. Without this explanation local fire officials may not be able to fully understand how the code can be applied and more importantly how it can be altered to fit a certain situation (Brooks 2011). For example, pathways between PV systems are required for commercial buildings. They are required from the outside perimeter access areas to ventilation areas such as skylights. Brooks (2011) states, "These pathways ensure that the fire fighter is free to move around the perimeter and to access the ventilation location closest to the fire." As such, if two ventilation areas are relatively close together, then these nearby ventilation areas may be considered one entity such that the requirements may not have to be duplicated in each ventilation area.

These requirements for PV systems on roofs are addressed in Section 11.12 Photovoltaic Systems of NFPA 1. These provisions include marking requirements for the main service disconnect, circuit disconnecting means, conduit and cable assembly, secondary power source, and inverters. Provisions for access, pathways, and smoke ventilation are included. Minimum distances of the array from the edges of the roof are described so that a fire fighter will have room to properly access the fire safely. The locations of the Direct Current (DC) Conductors are also detailed to provide maximum ventilation opportunities and minimize trip hazards. These are the most current adopted requirements that should be used in the installation of PV systems.

Important hazard mitigation measures have been incorporated into the 2014 NEC®, addressing fire fighter safety concerns that arose from the Target fire in Bakersfield, as well as some other incidents. One of the key changes in the NEC is a new section, 690.12, addressing rapid shutdown:

690.12 Rapid Shutdown of PV Systems on Buildings.
PV system circuits installed on or in buildings shall include a rapid shutdown function that controls specific conductors in accordance with 690.12(1) through (5) as follows.

(1) Requirements for controlled conductors shall apply only to PV system conductors of more than 1.5 m (5 ft) in length inside a building, or more than 3 m (10 ft) from a PV array.
(2) Controlled conductors shall be limited to not more than 30 volts and 240 volt-amperes within 10 seconds of rapid shutdown initiation.
(3) Voltage and power shall be measured between any two conductors and between any conductor and ground.
(4) The rapid shutdown initiation methods shall be labeled in accordance with 690.56(B).
(5) Equipment that performs the rapid shutdown shall be listed and identified (NEC 2014).

The NEC does not specify which equipment must perform the rapid disconnect function. It could be at the string level, at a module level dc–dc converter (which may be a power optimizer), at single module micro inverter, or at the module itself.

This leads into the key issue of PV systems, which is that they are almost always active. Many more provisions are described in Article 690 and 705 of the NEC. Provisions for marking, pathways, and smoke ventilation areas are described so that fire fighters can safely control the fire. The best practice is to use these articles along with NFPA 1, and all of their requirements should be met.

11.9 Weather-Related Maintenance Considerations

The rating of the roof is affected during both the installation of the PV systems as well as throughout the lifetime of the PV system. If the PV system is not installed correctly, this can lead to severe maintenance problems. Tampering with the roof structure can cause wear and tear over time and could further decrease the rating of the roof. In response to these questions, The Center PV Taskforce created a document that identifies four main objectives when installing PV panels in order to minimize their negative impact on the roof.

The following four objectives cover different hazards that are introduced to the roof when installing a PV system:

- PV module attachments and workers should not compromise the roof's water proofing integrity, especially through transferring loads in a way that avoids overloading or damaging the roof system.
- Every part connected to or used by the PV module should be designed for performance, durability and service life equivalent to the expected PV module service life. The entire system could be compromised if any element does not meet the criteria used for the entire PV system.
- Water needs to be able to drain correctly from the roof during and after PV installation. Siting pools of water can degrade roofing and the systems components as well.
- Provide accessibility to the roof and PV system design to allow for effective inspection, maintenance and repair.

By following these guidelines, PV systems and the roofs they are installed on can keep their fire resistance integrity, therefore protecting the building they are on and maximizing their production potential.

Chapter 12
Hazard Gap Analysis

12.1 General Knowledge Gaps

Limited documentation on all incident data for snow, wind, hail, fire, etc. has inhibited quantifying the risk associated with PV systems. Proposing effective solutions to problems can only be provided with data or information about past events. Although some data is known for fire incidents, the information is incomplete. It would be useful to compile estimated property losses and loss of life for each incident involving PV systems as well as the cumulative estimated total property loss and loss of life for an entire year. Also, little is known about the causes of an incident, e.g. failure in the panel, wiring, or attachment method and about factors contributing to fire spread during an incident.

The relative newness of rooftop PV systems limits the knowledge of longevity. "The manufacturer warranty period typically exceeds 20 years for crystalline silicon modules and 15 years for thin-film modules. Unfortunately, there is little or no systematically field monitored data or independent accelerated test data available to support most of these warranty claims. Investors, financiers, power purchasing agreement companies, and consumers are now expecting objective substantiations for these warranty claims. The PV module components, including cells and polymeric materials, must be protected from degradative losses (soft/durability losses) and catastrophic failures (hard/reliability failures) caused by stresses including temperature, humidity, ultraviolet (UV) radiation, wind, hail, and high system voltages, as well as effects including corrosion, broken interconnects, hotspots, delamination, and encapsulant discoloration" (TamizhMani and Kuitche 2013).

© Fire Protection Research Foundation 2015
R. Wills et al., *Best Practices for Commercial Roof-Mounted Photovoltaic System Installation*, SpringerBriefs in Fire, DOI 10.1007/978-1-4939-2883-5_12

12.2 Testing and Calculation Gaps

General testing limitations for PV systems are prevalent because each system is unique and is further challenged by the continuing development of new products and a wide variety of system designs. In addition, each installed PV system will be exposed to a unique environment. Current testing standards cannot address failure mechanisms for all climates and system integrations. Because PV systems are expensive, it is infeasible to test multiple scenarios. Testing only exposes PV systems to short term effects and cannot properly represent the degradative effects over the lifetime of a PV system. Therefore these testing methods cannot fully quantify the expected lifetime for the intended application/climate.

12.3 Structural and Wind Loads

Although SEAOC PV2-2012 may currently be the most effective way to calculate wind forces, it still has its limitations. Structurally attached and ballast-only PV systems are the only attachment method that has a determined way of predicting wind forces. Many PV systems are not structurally attached and need an alternate method to predict wind forces. A few of these attachment methods include ballast-only, screws, clips, fasteners, or adhesives, and anchoring to the roof deck. ASCE 7 discusses how to calculate wind forces using these attachment methods but the discussion is limited. Also, deflectors and shrouds are not addressed within SEAOC Report PV2-2012.

12.4 Hail Impact Forces

The current testing purpose for hail exposure is to assess short term effects and does not identify and quantify long term effects. There is no information on field degradation data versus accelerated test data. "Although these testing standards are helpful in creating PV systems that can withstand early infant mortality rate, there are limitations to these tests. These tests do not identify and quantify wear-out mechanisms. They do not address failure mechanisms for all climates and system integrations. Finally these tests are not meant to quantify the lifetime for the intended application/climate" (Wohlgemuth 2012). Hail impact is a very simple mechanical test and was never intended to try to determine long term effects of hail. It is a benchmark test to assess whether a PV module can withstand the vast majority of hail storms that a PV module will ever encounter. In that purpose it has been a highly successful test and failures of this test are rare. Hail is one of the hazards that affect the weathering of a PV system which is discussed in the general knowledge gaps.

12.5 Snow Loads

In order to provide a complete assessment, snow loads could be calculated in a similar manner as wind loads. However, documentation of this type of analysis is limited. Snow loads are addressed in qualification tests such as IEC 61215 for the module itself. An accelerated stress test using static mechanical loading for PV systems represents loads that snow buildup would introduce. Other than these few resources, there is little information on interaction of snow and PV system.

12.6 Seismic Hazards

Testing is the main method used to determine the seismic hazards associated with rooftop PV systems. Although these testing methods have been developed to best represent the same forces that will be exerted on to a PV system and its roof, there are still limitations to these testing methods. One limitation is representing the behavior of PV elements on top of a frozen roof along with seismic hazards.

> For solar arrays on buildings assigned to Seismic Design Category D, E, or F where rooftops are subject to significant potential for frost or ice that is likely to reduce friction between the solar array and the roof, the building official at their discretion may require increased minimum separation, further analysis, or attachment to the roof (SEAOC Report PV1-2012).
> The PV Committee is not aware of any research specifically addressing (a) the potential for frost or freezing of this type, (b) the effect of frost on the friction behavior of various roof surfaces, or (c) the likelihood that such frost forms underneath or sufficiently adjacent to solar panel feet as to compromise displacement resistance. Section C10.2 of ASCE 7-10 describes some of the phenomena related to the formation of frost, freezing rain, and ice (SEAOC Report PV1-2012).

The behavior of PV systems under seismic loads has been studied but there are still limitations on understanding the how each unique assembly is affected. This is also a limitation for many of the other hazards discussed in this report.

12.7 Fire Hazards

The 2012 edition of the IBC requires the PV system to be of the same fire class rating as the roof. UL 1703 has been recently updated to address testing methods to achieve this overall rating. The changes in the testing standard are so new that very few products have been evaluated to the new standard. It will take some time for products to be produced and evaluated to the new test standard so that PV systems can comply with building code requirements as they are implemented. Another main limitation with this is that PV systems can be so unique that there is no simple way to determine the fire resistance rating. Currently almost every possible combination of PV system and roof would have to be tested. Testing needs to be reasonable or else a drop in PV system installations would be expected, particularly in areas that require the highest fire class ratings.

The development of NFPA 1 and the NEC, has brought many new provisions that limit the possibly of an electrical failure of a PV system. A significant number of changes have been incorporated into these two codes to recent concerns involving PV systems. These new requirements for marking, pathways, and ventilation areas have increased the fire fighters' ability to control a fire safely. Further developments will need to be made to make sure these requirements are specific enough to apply to all circumstances.

Within these codes further requirements have been made regarding blind spots within ground fault detection. Is not possible to provide complete detection for both faults within the array and faults to the grounded CCC using a fuse, where optimizes this process but does not give a complete solution (Flicker and Johnson 2013).

References

Andrews RW, Pearce JM (2012) Prediction of energy effects on photovoltaic systems due to snowfall events, photovoltaic specialists conference (PVSC). In: 38th IEEE. http://www.academia.edu/1991659/Prediction_of_Energy_Effects_on_Photovoltaic_Systems_due_to_Snowfall_Events

Arndt and Puto (2011) Basic understanding of IEC standard testing for photovoltaic panels. http://tuvamerica.com/services/photovoltaics/ArticleBasicUnderstandingPV.pdf

Article 690, NFPA 70® (2008) National Electrical Code®, National Fire Protection Association, Quincy, MA, 2008 edition

ASTM E1038 (2010) Standard test method for determining resistance of photovoltaic modules to hail by impact with propelled ice balls, ASTM, International

Backstrom B, Tabaddor M (2010) PhD, effect of rack mounted photovoltaic modules on the fire classification rating of roofing assemblies. http://www.solarabcs.org/current-issues/docs/UL_Report_PV_Roof_Flammability_Experiments_11-30-10.pdf. Accessed 20 Nov

Backstrom R, Dini D (2011) Fire fighter safety and photovoltaic installations research project, underwriters laboratories, Inc., November 29. http://www.ul.com/global/documents/offerings/industries/buildingmaterials/fireservice/PV-FF_SafetyFinalReport.pdf

Backstrom B, Sloan D (2012) Effect of rack mounted photovoltaic modules on the fire classification rating of roofing assemblies phase 2, UL LLC, project number: 10CA49953, file number: IN15911, January 30. http://www.solarabcs.org/current-issues/docs/UL_Report_Phase2_1-30-12.pdf

Ball G (2011) Grounding photovoltaic modules: the lay of the land, solar ABCs interim report, solar America board for codes and standards, March. http://www.solarabcs.org/about/publications/reports/module-grounding/ pdfs/module-grounding_studyreport.pdf

Ball G, Zgonena T, Flueckiger C (2012) Photovoltaic module grounding: issues and recommendations, solar ABCs report, solar America board for codes and standards, April. http://www.solarabcs.org/about/publications/reports/modulegrounding/pdfs/IssuesRecomm_Grounding2_studyreport.pdf

Barkaszi S, O'Brien C (2010) Solar American board for codes and standards report: wind load calculations for PV arrays, June. http://www.solarabcs.org/about/publications/reports/windload/pdfs/Wind_Load_blanksstudyreport3.pdf

Brooks B (2011) Understanding the cal fire solar photovoltaic installation guideline, solar America board for codes and standards, March. http://solarabcs.com/about/publications/reports/fireguideline/pdfs/CslFire_studyreport.pdf

Brooks B (2012) The ground-fault protection BLIND SPOT: a safety concern for larger photovoltaic systems in the United States, a solar ABCs white paper, January. http://www.solarabcs.org/about/publications/reports/blindspot/pdfs/BlindSpot.pdf

Brooks B (2013) Private communication, October

CALFIRE (2008) Solar photovoltaic installation guideline, April 22. http://osfm.fire.ca.gov/pdf/reports/solarphotovoltaicguideline.pdf

© Fire Protection Research Foundation 2015

R. Wills et al., *Best Practices for Commercial Roof-Mounted Photovoltaic System Installation*, SpringerBriefs in Fire, DOI 10.1007/978-1-4939-2883-5

CALFIRE (2010) CALFIRE—office of the state fire marshal, fire operations for photovoltaic emergencies, November. (http://osfm.fire.ca.gov/training/pdf/Photovoltaics/Fire%20Ops% 20PV%20lo%20resl.pdf)

Davidson J, Orner F (2008) The new solar electric home: the complete guide to photovoltaics for your home

DiSanto (2013) Crews battle fire at christmas warehouse. http://www.nbcphiladelphia.com/news/ local/Fire-Burns-at-Christmas-Warehouse-233937231.html

Duval R (2013) Private communication, October 25

EPRI (2010) Addressing solar photovoltaic operations and maintenance challenges: a survey of current knowledge and practices, electric power research institute, July. http://www. smartgridnews.com/artman/uploads/1/1021496AddressingPVOaMChallenges7-2010_1_.pdf

Fire Engineering (2013) 11-alarm fire destroys south jersey warehouse, September. http://www. fireengineering.com/articles/2013/08/11-alarm-fire-destroys-south-jersey-warehouse.html

Flicker J, Johnson J (2013) Analysis of fuses for 'blind spot' ground detection in photovoltaic power systems

FOX (2013) Dietz and Watson fire aftermath, September. http://www.myfoxphilly.com/story/ 23326445/crews-still-pouring-water-on-dietz-watson-fire-scene)

FM 4470 (2012) Approval standard for single-ply, polymer-modified bitumen sheet, built-up roof (BUR) and liquid applied roof assemblies for use in class 1 and noncombustible roof deck construction. FM 4470, FM Approvals LLC, June. http://www.fmglobal.com/assets/pdf/ fmapprovals/4470.pdf

FM 4473 (2005) Specification test standard for impact resistance testing of ridgid roofing materials by impacting with freezer ice balls, FM 4473, Approvals LLC, July. http://www.fmglobal.com/ assets/pdf/fmapprovals/4473.pdf

FM 4476 (2011) Approval standard for flexible photovoltaic modules, FM 4476, Approvals LLC, January. http://www.fmglobal.com/assets/pdf/fmapprovals/4476.pdf

FM 4478 (2012) Approval standard for rigid photovoltaic modules, FM 4478, FM Approvals LLC, March. http://www.fmglobal.com/assets/pdf/fmapprovals/4478.pdf

Gandhi P (2011) Critical research for the success of photovoltaic systems. Presented to India Fire Advisory Council, UL, September 21

Grant CC (2010) Fire fighter safety and emergency response for solar power systems, final report, fire protection research foundation, May. http://www.nfpa.org/assets/files/pdf/research/ fftacticssolarpower.pdf

ICC ES (2012) Acceptance criteria for modular framing systems used to support photovoltaic (PV) panels. AC428, international code council evaluation service, November. http://www.icc-es. org/criteria/pdf_files/AC428.pdf

Jenkins DR, Mathey RG (1992) Hail impact testing procedure for solar covers. NBSIR 82-2487, National bureau of standards, April. Available from national technical information service, 285 Port Royal Road, Springfield, VA 22161-0001

Kopp G, Maffei J, Tilley C (2011) Rooftop solar arrays and wind loading: a primer on using wind tunnel testing as a basis for code compliant design per ASCE 7. Boundary layer wind tunnel laboratory, the university of western Ontario, faculty of engineering. http://www.sunlink.com/ files/WindPrimer_07-2011.pdf

Kopp (2012) Wind loads on low profile, tilted, solar array placed on low-rise buildings roofs. ATC/SEI hurricane conference, Miami

Kopp (2014) Wind loads on low profile, tilted, solar arrays placed on large, flat, low-rise building roofs. J Struct Eng. doi:10.1061/(ASCE)ST.1943-541X.0000821

Kopp, Banks (2012) Use of the wind tunnel test method for obtaining design wind loads on roof-mounted solar arrays. J Struct Eng

Kopp GA, Farquhar S, Morrison MJ (2013) Aerodynamic mechanisms for wind loads on tilted, roof-mounted, solar arrays. J Wind Eng Ind Aerodyn 40–52

Kirby JR (2011) Low-slope roofs as platforms for PV systems. SOLARPRO 32–40

Maffei J, Telleen K, Ward R, Kopp G, Schellenberg A (2014) Wind design practice and recommendations for solar arrays on low-slope roofs. J Struct Eng 140(2):04013040

Moore D, Wilson A (1978) Photovoltaic solar panel resistance to simulated hail. Low-cost solar array project report 5101-62, in national fire protection association, 2011: the U.S. fire problem, Quincy, MA. www.nfpa.org/categoryList.asp?categoryID=953&URL=Research/Fire %20statistics/T he%20U.S.% 20fire%20problem

NEC (2014) The national electrical code. National fire protection association

News NJ (2013) Deitz & Watson warehouse blaze, September. http://www.nj.com/burlington/ index.ssf/2013/09/dietz_and_watson_warehouse_fire_solar_panels_make_battling_blaze_ much_harder_officials_say.html

O'Brien PE, C, Barkaszi PE, S (2010) Wind load calculations for PV arrays. Solar American board for codes and standarts report. Jet propulsion laboratory, Pasadena, CA, 1978. Available from national technical information service, 5285 Port Royal Road, Springfield, VA 22161-0001

Paiss (2011a) Part 1, solar photovoltaic safety for fire fighters. http://www.youtube.com/watch?v= X1GXF8iQnyY

Paiss (2011b) Part 2, solar photovoltaic safety for fire fighters. http://www.youtube.com/watch?v= cJsvkZj0oscQ

Ross MMD (1995) Snow and ice accumulation on photovoltaic arrays: an assessment of the TN conseil passive melting technology, report # EDRL 95-68 (TR), energy diversification research laboratory, CANMET, natural resources Canada, Varennes, September, 273 p. http://www. rerinfo.ca/documents/trPVSnowandRime.pdf

Schellenberg A, Maffei J, Telleen K, Ward R (2013) Structural analysis and applications of wind loads to solar arrays. J Wind Eng Ind, Aero

SEIA (2013) U.S. market installs 832 megawatts in Q2 2013; grows 15% over last quarter. Retrieved from SEIA. http://www.seia.org/research-resources/solar-industry-data

SEAOC (2012a) Structural seismic requirements and commentary for rooftop solar photovoltaic arrays. Structural engineers association of California, solar photovoltaic systems committee, report SEAOC PV1-2012, August

SEAOC (2012b) Wind design for low-profile solar photovoltaic arrays on flat roofs. Structural engineers association of California, solar photovoltaic systems committee, report SEAOC PV2-2012, August

Sherwood L, Backstrom B, Sloan D, Flueckiger C, Brooks B, Rosenthal A (2013) Fire classification rating testing of stand-off mounted photovoltaic modules and systems, August. http://www.solarabcs.org/about/publications/reports/flammability- testing/pdfs/SolarABCs-36-2013-1.pdf

Smith J (2011) Targeting safety in photovoltaic system installation and maintenance, Fluke application note. http://support.fluke.com/find-sales/Download/Asset/3977708_6003_ENG_ A_W.PDF

Solar ABCs (2011) Impacts on photovoltaic installations of changes to the 2012 international codes, a solar ABCs white paper, May. http://www.solarabcs.org/about/publications/reports/ 2012Codes/pdfs/ABCS-22whitepaper.pdf

SunLink (2012) SunLink seismic testing roof integrity report. http://www.sunlink.com/ uploadedFiles/Sunlink/Content/Files/Report_SeismicTestingRoofIntegrity_04-2012.pdf

The Center PV Taskforce (2013) PV racking and attachment criteria for effective low slope metal panel roof system integration. The Center PV Taskforce Washington, DC

Thomson Reuters (2013) Techstreet store. http://www.techstreet.com/products/1235145

Uselton RB (2012) Senior principal engineer, applied research group, lennox industries Inc. Construction of a hail gun for solar PV module testing. February 28, 2012. http://www1.eere. energy.gov/solar/pdfs/pvmrw12_poster_si_uselton.pdf

Wiles J, Jr (2012) Photovoltaic system grounding, solar America board for codes and standards, October 2012. http://www.solarabcs.org/about/publications/reports/systemgrounding/pdfs/ SystemGrounding_studyreport.pdf

Wohlgemuth (2012) IEC 61215: what it is and isn't. National renewable energy laboratory. http://www.nrel.gov/docs/fy12osti/54714.pdf

XWOW (2013) Organic valley headquarters fire. News 19, La Crosse, WI. http://www.wxow.com/story/22251772/2013/05/14/fire-at-organic-valley-hq-in-la-farge

Printed in the United States
By Bookmasters